STAGE 1

2

Alan McSeveny Rachel McSeveny Diane McSeveny-Foster

Pearson Australia
(a division of Pearson Australia Group Pty Ltd)
459–471 Church St, Level 1, Building B, Richmond, Victoria, 3121
PO Box 23360, Melbourne, Victoria 8012
www.pearson.com.au

First published 2023 by Pearson Australia
2026 2026 2025 2024
10 9 8 7 6 5 4 3 2 1

Publishers: Sophie Matta and Rachel Elliott
Project Manager: Michelle Thomas
Production Editor: Laura Rentsch
Editor: Katie Millar
Designer: Anne Donald
Proofreader: Laura Rentsch
Rights & Permissions Editor: Alice McBroom
Cover Design: Jennifer Johnston
Cover Art: Michael Barter
Illustrator: Michael Barter
Publishing Services: Jit-Pin Chong
Printed in Malaysia (CTP-VVP)

National Library of Australia Cataloguing-in-Publication entry

A catalogue record for this book is available from the National Library of Australia

ISBN 978 0 6557 0903 9
Pearson Australia Group Pty Ltd ABN 40 004 245 943

Acknowledgement of Country
Pearson respects and honours Aboriginal and Torres Strait Islander Elders past, present and future. We acknowledge the stories, traditions and living cultures of the Traditional Custodians of the lands on which our company is located and where we conduct our business. Pearson is committed to honouring Australian Aboriginal and Torres Strait Islander peoples' unique cultural and spiritual relationships to the land, waters and seas and their rich contribution to society.

Aboriginal and Torres Strait Islander peoples are advised that this text may contain images, voices and names of deceased persons.

Attributions:
Shutterstock: Mega Pixel, p. 105.

What is Australian Signpost Maths NSW?

Australian Signpost Maths NSW is a mathematics program providing direction and support for teaching and learning. The series covers the content and skills presented in the NSW Mathematics Syllabus K–6, 2022.

A Student Book and an online Teacher Resource are provided for Kindergarten (Early Stage 1).

For Years 1 to 6 (Stages 1–3), a Student Book, an online Teacher Resource and a Mentals Book are provided for each year level. The online Teacher Resources provide a wealth of support for teachers.

The content has been carefully sequenced within each year level and across the K–6 series to take into account students' expected mathematical development. However, from the rich and varied material provided, teachers can develop individual learning programs to meet the needs of each student.

The Student Books are designed to support explicit teaching methods. Many group activities are provided in Activity, Investigation and Fun spots within the Student Books and the online Teacher Resource.

To maximise the benefits of the program, the Student Book, the online Teacher Resource and the Mentals Book should be used together.

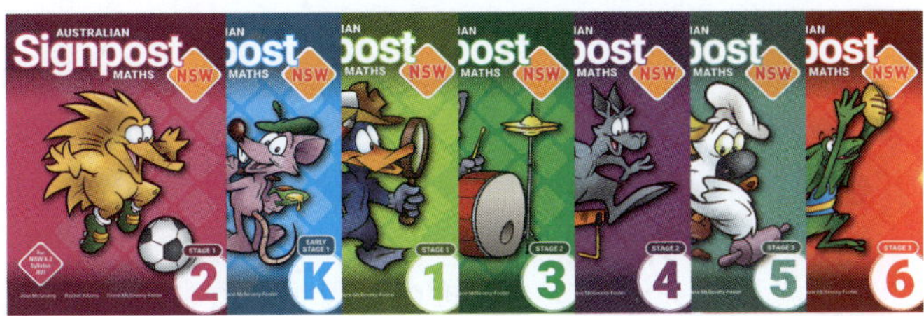

Student Books

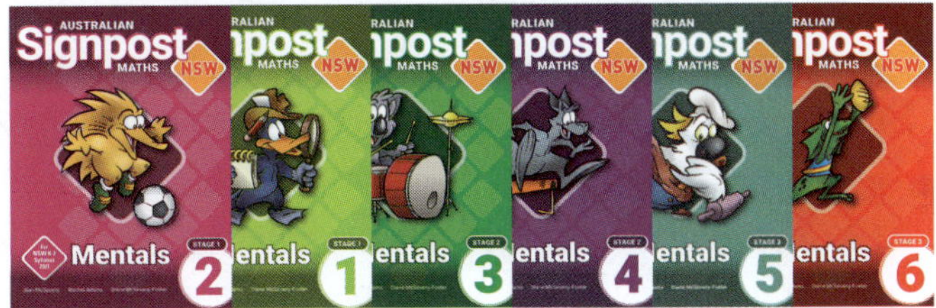

Mentals Books

Teacher Resource

Structure of Australian Signpost Maths NSW

In the K–2 books, the worksheet pages covering all three strands are presented in a recommended order. Each unit of 4 pages usually begins with Number and algebra. The Contents cross-reference allows teachers to quickly find the pages where each concept has been covered.

Within the program, explicit teaching, working mathematically skills, language development and identification and treatment of weaknesses are given high priority.

Identification and addressing areas of need

Five progress tests are designed to identify each student's areas of need, and the follow-up program after each of the tests is designed to address these needs. A reference to the relevant worksheet page is given for each test question. A remediation record page is used to track the student's progress.

These testing resources can be found in the online Teacher Resource.

Parallel progress retests are provided for further testing after remediation has taken place. See pages 142 and 143 of this book for more information.

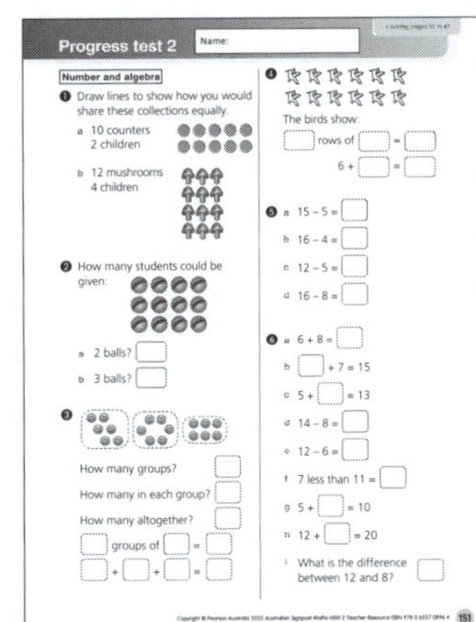

Progress test 2 Name:

Number and algebra

1 Draw lines to show how you would share these collections equally.
a 10 counters 2 children
b 12 mushrooms 4 children

2 How many students could be given:
a 2 balls? ☐
b 3 balls? ☐

3 How many groups? ☐
How many in each group? ☐
How many altogether? ☐
☐ groups of ☐ = ☐
☐ + ☐ + ☐ = ☐

4 The birds show:
☐ rows of ☐ = ☐
6 + ☐ = ☐

5 a 15 – 5 = ☐
b 16 – 4 = ☐
c 12 – 5 = ☐
d 16 – 8 = ☐

6 a 6 + 8 = ☐
b ☐ + 7 = 15
c 5 + ☐ = 13
d 14 – 8 = ☐
e 12 – 6 = ☐
f 7 less than 11 = ☐
g 5 + ☐ = 10
h 12 + ☐ = 20
i What is the difference between 12 and 8? ☐

151

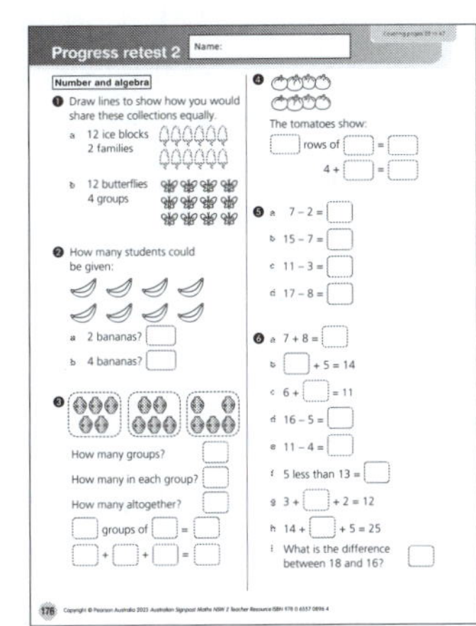

Progress retest 2 Name:

Number and algebra

1 Draw lines to show how you would share these collections equally.
a 12 ice blocks 2 families
b 12 butterflies 4 groups

2 How many students could be given:
a 2 bananas? ☐
b 4 bananas? ☐

3 How many groups? ☐
How many in each group? ☐
How many altogether? ☐
☐ groups of ☐ = ☐
☐ + ☐ + ☐ = ☐

4 The tomatoes show:
☐ rows of ☐ = ☐
4 + ☐ = ☐

5 a 7 – 2 = ☐
b 15 – 7 = ☐
c 11 – 3 = ☐
d 17 – 8 = ☐

6 a 7 + 8 = ☐
b ☐ + 5 = 14
c 6 + ☐ = 11
d 16 – 5 = ☐
e 11 – 4 = ☐
f 5 less than 13 = ☐
g 3 + ☐ + 2 = 12
h 14 + ☐ + 5 = 25
i What is the difference between 18 and 16? ☐

176

Special features of Australian Signpost Maths NSW

- **The traffic light icons**
  These are found on the top right of each worksheet page in the Student Books. They allow students to assess their own progress and give feedback to the teacher.
  - ☐ **Green:** I found this work easy.
  - ☐ **Orange:** I found some work on the page difficult.
  - ☐ **Red:** I don't understand the work on this page.

- **Dictionary**
  Terms used in the Student Book and terms that should be understood at this level are recorded here to provide a reference for students and teachers. This is found on pages xiii–xviii of this book and in the online Teacher Resource.

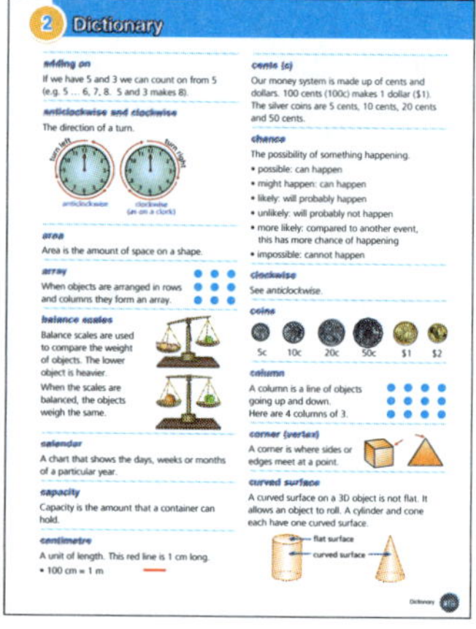

2 Dictionary

adding on
If we have 5 and 3 we can count on from 5 (e.g. 5 … 6, 7, 8. 5 and 3 makes 8).

anticlockwise and clockwise
The direction of a turn.

area
Area is the amount of space on a shape.

array
When objects are arranged in rows and columns they form an array.

balance scales
Balance scales are used to compare the weight of objects. The lower object is heavier.
When the scales are balanced, the objects weigh the same.

calendar
A chart that shows the days, weeks or months of a particular year.

capacity
Capacity is the amount that a container can hold.

centimetre
A unit of length. This red line is 1 cm long.
• 100 cm = 1 m

cents (c)
Our money system is made up of cents and dollars. 100 cents (100c) makes 1 dollar ($1). The silver coins are 5 cents, 10 cents, 20 cents and 50 cents.

chance
The possibility of something happening.
• possible: can happen
• might happen: can happen
• likely: will probably happen
• unlikely: will probably not happen
• more likely: compared to another event, this has more chance of happening
• impossible: cannot happen

clockwise
See anticlockwise.

coins

column
A column is a line of objects going up and down. Here are 4 columns of 3.

corner (vertex)
A corner is where sides or edges meet at a point.

curved surface
A curved surface on a 3D object is not flat. It allows an object to roll. A cylinder and cone each have one curved surface.

Dictionary

- **ID cards (Years 1 to 6)**
  These cards review the language of Mathematics by asking students to identify common terms, shapes and symbols. They are designed to be reused and are found in the online Teacher Resource and in the front of the Mentals Books.

- **Progress tests**
  These allow the teacher to identify each student's strengths and needs. Cross-references for each question direct teachers and students to the pages where that work is introduced. Tables are provided to record the follow-up that takes place and parallel tests are provided for retesting. These tests can be found in the online Teacher Resource.

- **Year 2 Consolidation booklet**
  This 32 page booklet is found in the online Teacher Resource. It is designed to reinforce work completed in class and provides practice of important skills and addition and subtraction facts. The booklet can be used when there is limited supervision or when a student finishes classwork early.

- **Answers**
  These are supplied in the online Teacher Resource.

- **Blackline masters (BLM)**
  References are made to the blackline masters in the teaching suggestions provided for each student work page.

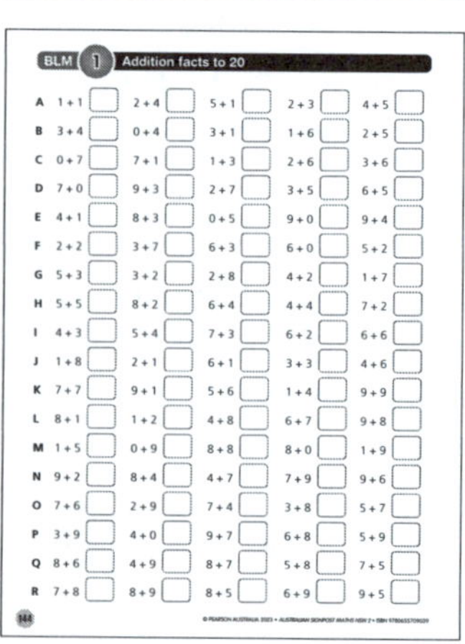

BLM 1 Addition facts to 20

A	1 + 1	2 + 4	5 + 1	2 + 3	4 + 5
B	3 + 4	0 + 4	3 + 1	1 + 6	2 + 5
C	0 + 7	7 + 1	1 + 3	2 + 6	3 + 6
D	7 + 0	9 + 3	2 + 7	3 + 5	6 + 5
E	4 + 1	8 + 3	0 + 5	9 + 0	9 + 4
F	2 + 2	3 + 7	6 + 3	6 + 0	5 + 2
G	5 + 3	3 + 2	2 + 8	4 + 2	1 + 7
H	5 + 5	8 + 2	6 + 4	4 + 4	7 + 2
I	4 + 3	5 + 4	7 + 3	6 + 2	6 + 6
J	1 + 8	2 + 1	6 + 1	3 + 3	4 + 6
K	7 + 7	9 + 1	5 + 6	1 + 4	9 + 9
L	8 + 1	1 + 2	4 + 8	6 + 7	9 + 8
M	1 + 5	0 + 9	8 + 8	8 + 0	1 + 9
N	9 + 2	8 + 4	4 + 7	7 + 9	9 + 6
O	7 + 6	2 + 9	7 + 4	3 + 8	5 + 7
P	3 + 9	4 + 0	9 + 7	6 + 8	5 + 9
Q	8 + 6	4 + 9	8 + 7	5 + 8	7 + 5
R	7 + 8	8 + 9	8 + 5	6 + 9	9 + 5

144

- **Differentiation**
  Each student work page has a Teacher Resource page to support it. Cross-references direct the teacher to pages where the concept is introduced and developed. These references may be from the Student Book for the previous year, the current year or the next year.

  The Teacher Resource support pages provide additional learning activities for students who need remediation or extension activities. The Blackline Masters provide activities to support students of various learning abilities.

- **Cartoons**
  Cartoons are used to motivate and instruct.

Australian Signpost Maths NSW icons

Signpost icons are used throughout the book as cues to the essential nature of exercises and activities, and as a guide to ways of engaging with them. These icons often indicate alternative or more concrete approaches to dealing with concepts.

This icon highlights **important rules and concepts** occurring throughout the book. It often appears with worked examples.

Investigations allow students to **explore and discover** maths concepts.

Activities provide **applications and enrichment**. These activities usually involve the use of concrete materials and partner or group work.

These enjoyable activities are used to **motivate and involve** students in mathematical pursuits. They usually involve games and puzzles.

Structure of New South Wales Mathematics K–6

The NSW Mathematics Syllabus content is presented in three strands:

1 Number and algebra
2 Measurement and space
3 Statistics and probability

Working mathematically pervades each of these strands.

The Mathematics syllabus can be found at:
https://curriculum.nsw.edu.au/learning-areas/mathematics/mathematics-k-10

Textbook structure
Within the Contents for Year 2, we show related pages using these categories:

Number and algebra	**Measurement and space**	**Statistics and probability**
Numbers	2D shapes / 3D objects	Data displays
Addition / subtraction	Length / area / mass	Chance
Sharing / grouping	Capacity / volume	
Patterns	Time / duration	
	Position	

2 Contents and syllabus overview

KEY

Colour	Strand
Blue	Number and algebra
Green	Measurement and space
Orange	Statistics and probability

Page	Unit	Title	Strand: Number / algebra	Strand: Measurement / space	Strand: Statistics / probability	Content area: Numbers	Content area: Addition / subtraction	Content area: Sharing / grouping	Content area: Patterns	Content area: 2D shapes / 3D objects	Content area: Length / area / mass	Content area: Capacity / volume	Content area: Time / duration	Content area: Position	Content area: Data displays	Content area: Chance
1	Thinking Skills		Working mathematically pervades each of the strands.													
2	1A	Combinations to 10	■				●									
3	1B	Subtraction to 10	■				●									
4	1C	Position words		■										●		
5	1D	Modelling numbers	■			●										
6	2A	Addition	■				●									
7	2B	Addition to 20	■				●									
8	2C	Addition to 20	■				●									
9	2D	Thinking about graphs			■										●	
10	3A	Doubling and near doubling	■				●									
11	3B	Sharing	■					●								
12	3C	Sharing	■					●								
13	3D	2D shapes		■						●						
14	4A	Subtraction	■				●									
15	4B	Subtraction to 20	■				●									
16	4C	Ordinal numbers and calendars		■									●			
17	4D	The calendar		■									●			
18	5A	Addition to 20	■				●									
19	5B	Addition by looking for tens	■				●									
20	5C	Looking at 3D objects		■						●						
21	5D	Describing 3D objects		■						●						

Progress test 1: Administer test (Teacher Resource, pages 147–149) then address weaknesses.

Page	Unit	Title	Strand: Number / algebra	Strand: Measurement / space	Strand: Statistics / probability	Content area: Numbers	Content area: Addition / subtraction	Content area: Sharing / grouping	Content area: Patterns	Content area: 2D shapes / 3D objects	Content area: Length / area / mass	Content area: Capacity / volume	Content area: Time / duration	Content area: Position	Content area: Data displays	Content area: Chance
22	6A	Sharing and grouping	■					●								
23	6B	Groups and rows	■					●								
24	6C	Revision of time		■									●			
25	6D	Estimating time passed		■									●			
26	7A	Groups and rows	■					●								
27	7B	Problem solving	■					●								
28	7C	Features of 2D shapes		■						●						
29	7D	Drawing 2D shapes		■						●						

KEY

Colour	Strand
Blue	Number and algebra
Green	Measurement and space
Orange	Statistics and probability

Page	Unit	Title	Strand	Number / algebra	Measurement / space	Statistics / probability	Content area	Numbers	Addition / subtraction	Sharing / grouping	Patterns	2D shapes / 3D objects	Length / area / mass	Capacity / volume	Time / duration	Position	Data displays	Chance
30	8A (blue)	Subtraction to 20		■					●									
31	8B (blue)	Differences		■					●									
32	8C (green)	Balance scales			■								●					
33	8D (green)	Comparing masses			■								●					
34	9A (blue)	Linking addition and subtraction		■					●									
35	9B (blue)	Linking addition and subtraction		■					●									
36	9C (green)	Informal units of length			■								●					
37	9D (green)	Informal units of length			■								●					
38	10A (blue)	How many more?		■					●									
39	10B (green)	Volume and capacity			■									●				
40	10C (orange)	Using graphs				■											●	
41	10D (orange)	Chance				■												●
42	11A (blue)	Adding 10s		■					●									
43	11B (blue)	Adding and subtracting 10s		■					●									
44	11C (green)	Ordering capacities			■									●				
45	11D (orange)	Lists, graphs and tables				■											●	
46	12A (blue)	Inverse operations		■					●									
47	12B (blue)	Australian money		■				●										

Progress test 2: Administer test (Teacher Resource, pages 151–154) then address weaknesses.

Page	Unit	Title	Strand	Number / algebra	Measurement / space	Statistics / probability	Content area	Numbers	Addition / subtraction	Sharing / grouping	Patterns	2D shapes / 3D objects	Length / area / mass	Capacity / volume	Time / duration	Position	Data displays	Chance
48	12C (blue)	Halves / quarters		■						●								
49	12D (blue)	Fractions of a group		■						●								
50	13A (blue)	Equal groups		■						●								
51	13B (blue)	Equal groups		■						●								
52	13C (green)	Analog time			■										●			
53	13D (green)	Analog time			■										●			
54	14A (blue)	Using skip counting		■						●								
55	14B (blue)	Number lines		■					●	●								
56	14C (green)	Digital time			■										●			
57	14D (green)	Analog time			■										●			
58	15A (blue)	Using arrays		■						●								
59	15B (blue)	Arrays		■						●								
60	15C (blue)	Problem solving		■						●								
61	15D (orange)	Chance				■												●
62	16A (blue)	Numbers to 150		■				●										
63	16B (blue)	Numbers to 1000		■				●										
64	16C (green)	Informal units of length			■								●					
65	16D (orange)	Telling the story from data				■											●	

KEY

■ (blue)	Number and algebra
■ (green)	Measurement and space
■ (orange)	Statistics and probability

Page	Unit	Title	Strand	Number / algebra	Measurement / space	Statistics / probability	Content area	Numbers	Addition / subtraction	Sharing / grouping	Patterns	2D shapes / 3D objects	Length / area / mass	Capacity / volume	Time / duration	Position	Data displays	Chance
66	17A	Numbers to 1000		■				●										
67	17B	Numbers to 1000		■				●										
68	17C	Informal units of length			■								●					
69	17D	The metre			■								●					
70	18A	Numbers to 1000		■				●										
71	18B	Number patterns		■							●							
72	18C	Centimetres			■								●					
73	18D	Measuring with centimetres			■								●					
74	19A	Number patterns		■							●							
75	19B	Counting by tens		■				●										
76	19C	Comparing areas			■								●					
77	19D	Area			■								●					
78	20A	Numbers		■				●										
79	20B	Rounding to the nearest ten		■				●										
80	20C	Area using informal units			■								●					
81	20D	Area of a rectangle			■								●					
82	21A	Value of coins		■					●									
83	21B	Money		■					●									

Progress test 3: Administer test (Teacher Resource, pages 156–160) then address weaknesses.

Page	Unit	Title	Strand	Number / algebra	Measurement / space	Statistics / probability	Content area	Numbers	Addition / subtraction	Sharing / grouping	Patterns	2D shapes / 3D objects	Length / area / mass	Capacity / volume	Time / duration	Position	Data displays	Chance
84	21C	Duration using hours			■										●			
85	21D	Duration using weeks			■										●			
86	22A	Amounts to $2		■					●									
87	22B	Value of coins		■					●									
88	22C	Prisms and cylinders			■							●						
89	22D	3D objects			■							●						
90	23A	Building to the next 10		■					●									
91	23B	Building to the next 10		■					●									
92	23C	Volume			■									●				
93	23D	Comparing volume			■									●				
94	24A	Split strategy (addition)		■					●									
95	24B	Split strategy (addition)		■					●									
96	24C	Ordering masses			■								●					
97	24D	Balance scales			■								●					
98	25A	Rounding to the nearest 100		■				●										
99	25B	Building to the next 10		■					●									
100	25C	Turning a shape			■							●						
101	25D	Turning shapes			■							●						

KEY

■ (blue)	Number and algebra
■ (green)	Measurement and space
■ (orange)	Statistics and probability

Page	Unit	Title	Strand	Number / algebra	Measurement / space	Statistics / probability	Content area	Numbers	Addition / subtraction	Sharing / grouping	Patterns	2D shapes / 3D objects	Length / area / mass	Capacity / volume	Time / duration	Position	Data displays	Chance
102	26A	Using rows		■						●								
103	26B	Using groups		■						●								
104	26C	Adding columns		■						●								
105	26D	The cube			■							●						
106	27A	Jump strategy (addition)		■					●									
107	27B	Jump strategy (subtraction)		■					●									
108	27C	Giving directions			■											●		

Progress test 4: Administer test (Teacher Resource, pages 162–164) then address weaknesses.

Page	Unit	Title	Strand	Number / algebra	Measurement / space	Statistics / probability	Content area	Numbers	Addition / subtraction	Sharing / grouping	Patterns	2D shapes / 3D objects	Length / area / mass	Capacity / volume	Time / duration	Position	Data displays	Chance
109	27D	Using tally marks				■											●	
110	28A	Jump strategy		■					●									
111	28B	Fractions of a group		■						●								
112	28C	Eighths of a length			■								●					
113	28D	Making graphs				■											●	
114	29A	Doubling and halving		■						●								
115	29B	Doubling and halving		■						●								
116	29C	Duration of time			■										●			
117	29D	Gathering data				■											●	
118	30A	Problem solving		■						●								
119	30B	Problem solving		■						●								
120	30C	Combine and separate shapes			■							●						
121	30D	Following instructions			■											●		
122	31A	Repeated subtraction		■						●								
123	31B	Fractions of a whole			■								●					
124	31C	Fractions			■								●					
125	31D	Metres and centimetres			■								●					
126	32A	Choosing a strategy		■					●									
127	32B	Choosing a strategy		■					●									
128	32C	Quarter turns			■							●						

Progress test 5: Administer test (Teacher Resource, pages 166–168) then address weaknesses.

KEY

■ (blue)	Number and algebra
■ (green)	Measurement and space
■ (orange)	Statistics and probability

Page	Unit	Title	Strand	Number / algebra	Measurement / space	Statistics / probability	Content area	Numbers	Addition / subtraction	Sharing / grouping	Patterns	2D shapes / 3D objects	Length / area / mass	Capacity / volume	Time / duration	Position	Data displays	Chance
129	32D	Half and quarter turns			■							●						
130	33A	Related problems		■					●									
131	33B	Inverse strategy, subtraction		■					●									
132	33C	Describing 3D objects			■							●						
133	33D	3D objects			■							●						
134	34A	Money		■					●									
135	34B	Possible outcomes				■												●
136	34C	Gather and organise data				■											●	
137	34D	Comparing objects			■								●	●				
138	35A	Giving directions			■											●		
139	35B	More shapes (extension)			■							●						
140	35C	Problem solving with addition		■					●									
141	35D	Problem solving with groups		■						●								
142	Identifying and addressing areas of need																	
144	**1** Addition facts to 20	**2** Subtraction facts to 10																
146	**3** Subtraction facts to 20	**4** Skip counting / number chart																

Contents cross-reference

Number and algebra

Measurement and space

1	Geometric measure	Pages
	Position: Simple maps, directions and position	4, 108, 121, 138
	Ordinal numbers	16, 17, 116
	Left and right	121, 138
	Length: Informal units of length	36, 37, 64, 68, 137
	Formal units of length, metre and centimetre	69, 72, 73, 125
	Halves, quarters and eighths	48, 49, 123, 124, 128
2	**Two-dimensional spatial structure**	**Pages**
	2D shapes: Sorting, describing and making 2D shapes	13, 28, 29, 120, 139
	Combine and separate 2D shapes	29, 120
	Orientation of shapes, turning shapes, patterns	100, 101, 128, 129, 139
	Area: Describing and comparing rectangular areas	76, 77, 80, 81, 120
3	**Three-dimensional spatial structure**	**Pages**
	3D objects: Describing and sorting 3D objects	20, 21, 88, 89, 105, 132, 133
	Volume: Describing and comparing volumes, informal units	92, 93
	Capacity, using informal units	39, 44
4	**Non-spatial measure**	**Pages**
	Mass: Describing and comparing masses of objects, equal-arm balance	32, 33, 96, 97, 137
	Time: Duration using units of time, calendar, hour, minute, second	16, 17, 25, 84, 85, 116
	Months and seasons	116
	Telling time on the half and quarter hour using analog and digital clocks	24, 52, 53, 56, 57, 128, 129

Statistics and probability

1	Data	Pages
	Collecting information, identifying questions and gathering data	40, 109, 113, 117, 136
	Creating and interpreting data displays	9, 40, 45, 65, 109, 113, 117, 136
2	**Chance**	**Pages**
	Identifying and describing activities involving chance	41, 61, 135

2 Dictionary

adding on

If we have 5 and 3 we can count on from 5 (e.g. 5 … 6, 7, 8. 5 and 3 makes 8).

anticlockwise and clockwise

The direction of a turn.

anticlockwise

clockwise
(as on a clock)

area

Area is the amount of space on a shape.

array

When objects are arranged in rows and columns they form an array.

balance scales

Balance scales are used to compare the weight of objects. The lower object is heavier.

When the scales are balanced, the objects weigh the same.

calendar

A chart that shows the days, weeks or months of a particular year.

capacity

Capacity is the amount that a container can hold.

centimetre

A unit of length. This red line is 1 cm long.

- 100 cm = 1 m

cents (c)

Our money system is made up of cents and dollars. 100 cents (100c) makes 1 dollar ($1). The silver coins are 5 cents, 10 cents, 20 cents and 50 cents.

chance

The possibility of something happening.

- possible: can happen
- might happen: can happen
- likely: will probably happen
- unlikely: will probably not happen
- more likely: compared to another event, this has more chance of happening
- impossible: cannot happen

clockwise

See *anticlockwise*.

coins

5c

10c

20c

50c

$1

$2

column

A column is a line of objects going up and down.
Here are 4 columns of 3.

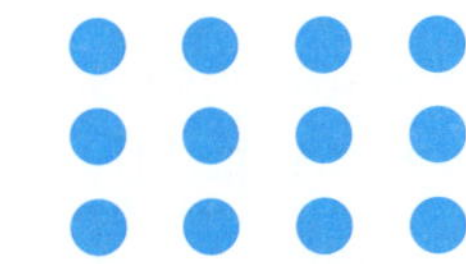

corner (vertex)

A corner is where sides or edges meet at a point.

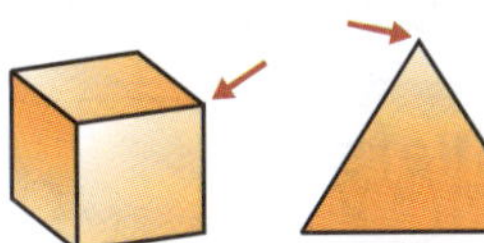

curved surface

A curved surface on a 3D object is not flat. It allows an object to roll. A cylinder and cone each have one curved surface.

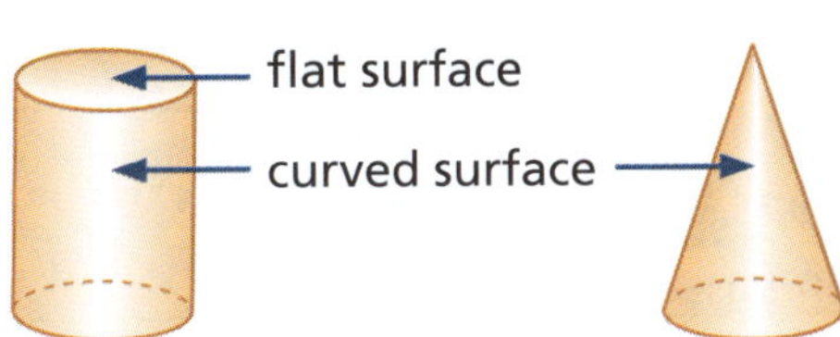

data display

A data display shows categories of objects and allows us to compare them.

Pets

Graph: Cats

Dogs

Table:

Dogs	Cats
4	2

date

The date shows the day, the month and the year. For example, 30.5.23 means the 30th day on the 5th month (May) in the year 2023.

difference

How many more?

The difference between 16 and 13 is 3.

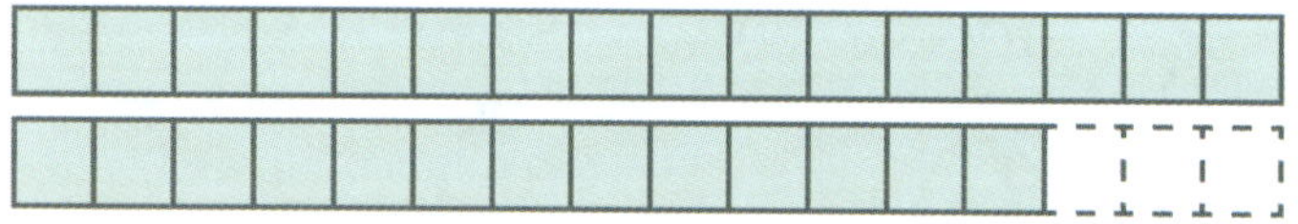

For smaller numbers, line up each group in a row to find the difference.

For larger numbers, place the numbers on a number line to find the difference.

dollars ($)

Our money system is made up of cents and dollars.

100 cents (100c) makes 1 dollar ($1).

The gold coins are $1 (1 dollar) and $2 (2 dollars).

The notes are $5, $10, $20, $50, $100.

double

Double means the same thing twice.
Double 4 means 4 + 4 = 8.

edge

Two faces of a 3D object meet at an edge.

eighth

One of eight equal parts.

One eighth of the rectangle is coloured.

equal groups

Groups that have the same number of members.

equals sign =

The equals sign means "makes", "is equal to" or "is the same as".

For example, 2 + 3 = 4 + 1.

estimate

A good guess.

face

A flat surface that has straight sides (e.g. the side of a box).

flat surface

A cylinder has 2 flat surfaces, one on both ends, and one curved surface.

A cube has 6 flat surfaces.

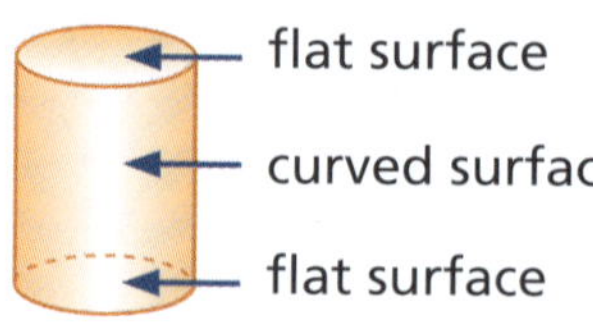

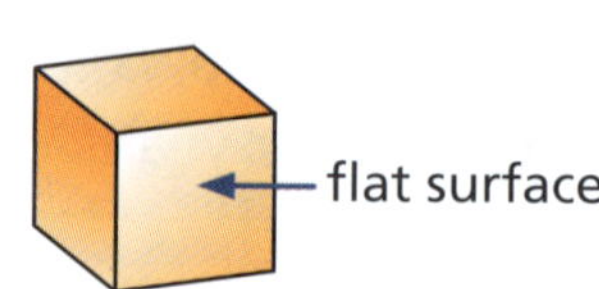

friends of ten

Numbers that add together to make 10.
The friends of 10 are 1 and 9, 2 and 8, 3 and 7, 4 and 6, 5 and 5, 6 and 4, 7 and 3, 8 and 2, 9 and 1.

graph

See *data display*.

half

One of two equal parts.

One half of the rectangle is coloured.

halfway point

The halfway point is the middle position.

hefting

To compare masses by lifting them with your hands.

hour

Hours are used to measure the length of time.

60 minutes is the same as one hour.

There are 24 hours in one day.

left and right

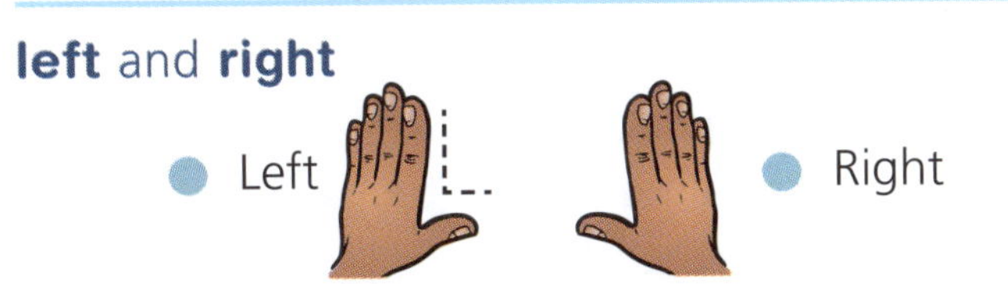

notes

number bonds

These show how a number can be broken up into parts (e.g. the top number, 4, can be broken up into 1 and 3).

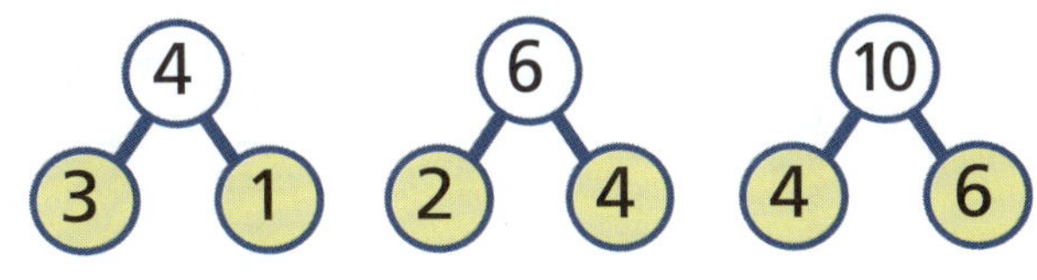

number line

A line that shows numbers in order. Number lines can be used for many things (e.g. counting, adding and subtracting).

number sentence

A number sentence can use words, numerals and symbols (e.g. 4 and 6 makes 10. This can be written as 4 + 6 = 10).

numeral

A numeral is a symbol that stands for a number such as 7, 18, 92, 120 or vii.

odd and even numbers

Odd numbers end in 1, 3, 5, 7 or 9 (e.g. 49).

Even numbers end in 2, 4, 6, 8 or 0 (e.g. 32).

ordinal numbers

Ordinal numbers describe the order or position of something.

1st, 2nd, 3rd, 4th, 5th …

pattern

A pattern is a group of numbers, objects, shapes, colours, sounds or actions that are repeated over and over again.

place value

The position of each digit of a numeral holds a different value.

place-value blocks

These are used to represent numbers.

ones block

tens block

hundreds block

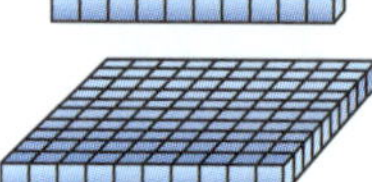

42 shown as

113 shown as

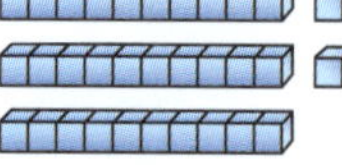
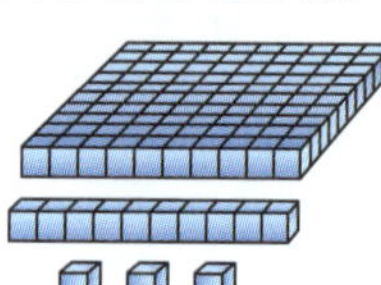

quarter

One of four equal parts.

One quarter of the rectangle is coloured.

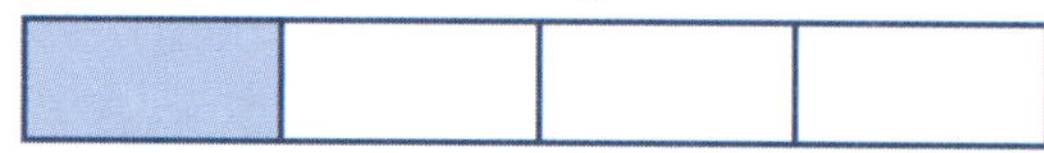

reflection (or flip)

A mirror image.

right and left

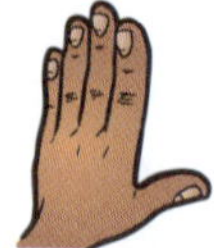
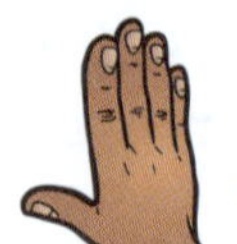

row

A row is a line of objects going across.

Here are 2 rows of 5.

second

Seconds are used to measure the length of time. 60 seconds is the same as 1 minute.

sharing

When sharing, we make sure that each share is the same size.

2 people could share these 6 balls. Each person would get 3 balls.

If two groups are not the same, we can make fair shares by moving items from the larger group to the smaller group.

slide

Moving a shape in any direction without changing the size or orientation.

straight line

A straight line has no bends or curves.

symmetry

A shape has symmetry when one side is the mirror-image of the other. It can be folded so that the two halves match, exactly.

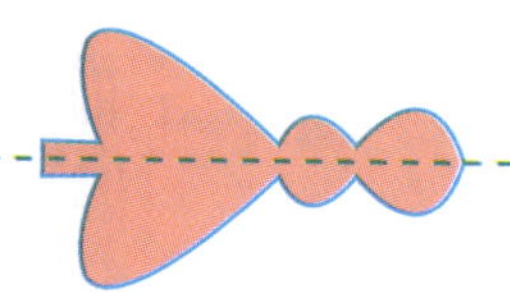

take away (subtract or minus)

When we remove objects from a group we call this "take away".

tally marks

Tally marks are used to keep count. Groups of 5 are used.

卌 = 5

three-dimensional (3D) objects

3D objects are three-dimensional.
They have length, width and height.

spheres (ball-shaped objects) are curved and round. They can roll.

cubes (box-shaped objects) have 6 square faces. They can slide.

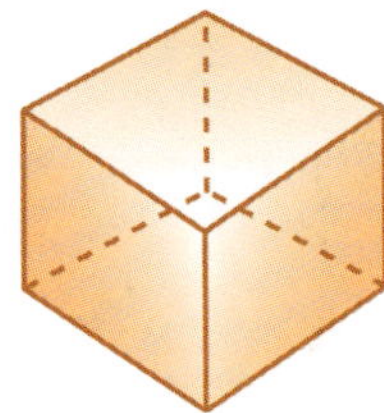

cylinders (can-shaped objects) have 2 flat surfaces and 1 curved surface. They can roll and slide.

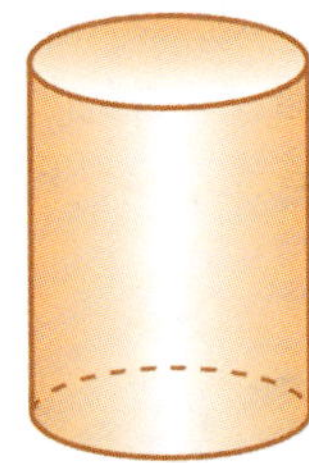

cones (cone-shaped objects) have 1 flat surface and 1 curved surface. They can roll and slide.

prisms

A prism has rectangular faces joining two identical bases at both ends.

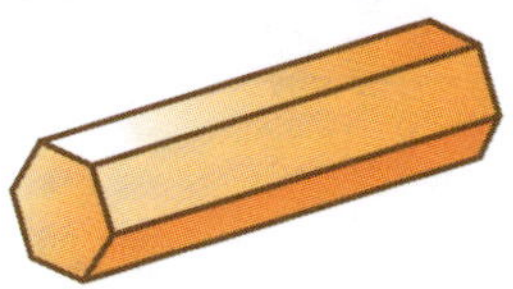

hexagonal prism

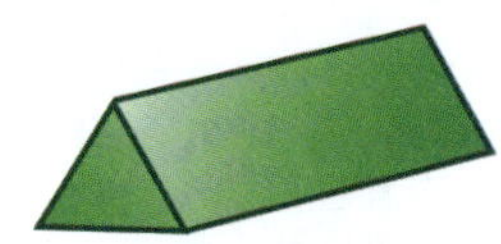

triangular prism

time words

morning	day
afternoon	night

Days			
Sunday	Monday	Tuesday	Wednesday
Thursday	Friday	Saturday	

Months			
January	February	March	April
May	June	July	August
September	October	November	December

Seasons			
Summer	Autumn	Winter	Spring

- clocks

analog clock

3 o'clock

digital clock

- o'clock

When the long hand (minute hand) is pointing to 12, the time is an "o'clock" time. The short hand (hour hand) points to the hour (e.g. the hour hand above is pointing to the 3 so it is 3 o'clock).

half past 3

- half past

When the long hand (hour hand) is pointing to the 6, the time is a "half past" time. The short hand (hour hand) above is pointing halfway between the 3 and the 4 so it is half past 3.

quarter past 6

- quarter past

When the long hand (minute hand) is pointing to the 3 and the short hand is a quarter of the way past the hour number, the time is a "quarter past" time.

quarter to 7

- quarter to

When the long hand (minute hand) is pointing to the 9 and the short hand is a quarter of the way from the next hour number, the time is a "quarter to" time.

total

The number of items altogether. The result once everything has been added.

turn

Moving a shape in a clockwise or anticlockwise direction, without changing the shape or size.

two-dimensional (2D) shapes

Flat shapes are two-dimensional.
They have length and width.

circle
1 curved side

triangle
3 sides
3 corners

square
4 equal sides
4 corners

rectangle
2 equal long sides and 2 equal short sides, like a stretched square

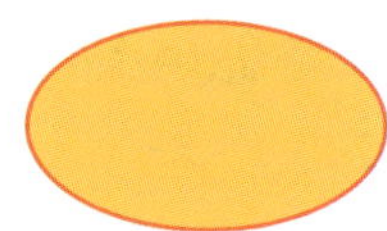

oval
1 curved side, like a squashed circle

pentagon
5 sides
5 corners

hexagon
6 sides
6 corners

octagon
8 sides 8 corners

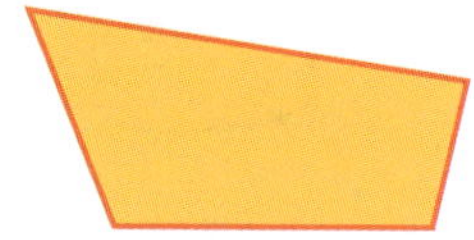

quadrilaterals
4 sides and 4 corners. Squares and rectangles are quadrilaterals.

vertex (plural is vertices)

A corner.

volume

The amount of space an object takes up.

year

There are 365 days in a year and 366 days in a leap year. There are 12 months in a year.

2 Thinking skills

Little robbers

1. What were the mice planning to do?
2. Do you think the mouse on stilts could reach the pie?
3. Explain how the mice carried out their plan.
4. Do you think the mice would take only a part of the pie? Why?
5. Does the police officer think the mice took the missing pie? Why or why not?
6. How do you think the stilts could be made?
7. Which of these questions do you like the best? Why do you like it?

1A Combinations to 10

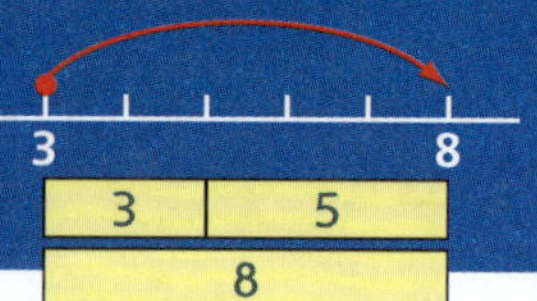

Emma has  

Paul has 

$$\begin{array}{r} \$3 \\ +\$5 \\ \hline \$8 \end{array}$$

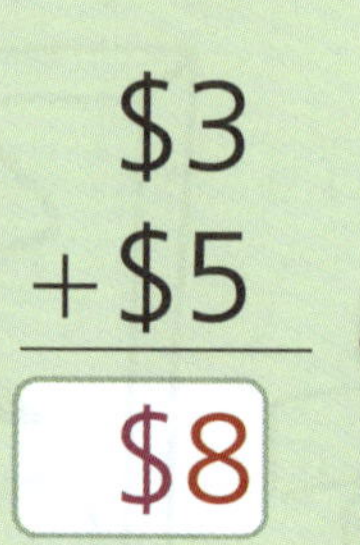

Altogether Emma and Paul have $8.

1 Count on from the largest number to complete these.

a	b	c	d	e
1 + 6 = ☐	2 + 2 = ☐	3 + 4 = ☐	1 + 3 = ☐	4 + 5 = ☐

f	g	h	i	j
4 + 4 = ☐	6 + 3 = ☐	5 + 4 = ☐	6 + 1 = ☐	0 + 5 = ☐

k	l	m	n	o
2 + 7 = ☐	5 + 1 = ☐	2 + 5 = ☐	3 + 6 = ☐	7 + 0 = ☐

Discuss word problems for these number sentences.

2 Finish the **number story** for each question below.

a   +  ☐ lollies + ☐ lollies equals ☐ lollies.

b  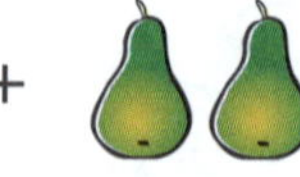 + ☐ pears + ☐ pears equals ☐ pears.

Use objects to model your own addition stories.

 • *AUSTRALIAN SIGNPOST MATHS NSW 2* • ISBN 9780655709039

1B Subtraction to 10

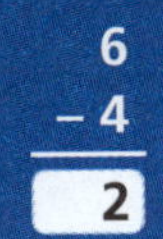

1 You could use counting on or counting back to complete these.

a 6 – 4 = ☐ **b** 8 – 3 = ☐ **c** 9 – 6 = ☐

d 10 – 7 = ☐ **e** 9 – 5 = ☐ **f** 10 – 8 = ☐

g 7 – 4 = ☐ **h** 10 – 4 = ☐ **i** 9 – 3 = ☐

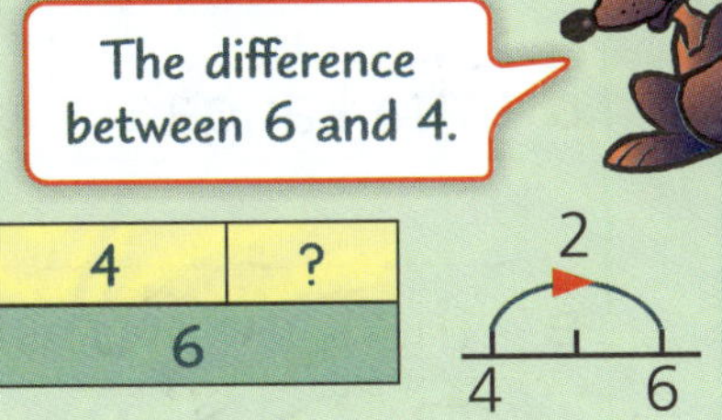

2

a 5 – 3 = ☐ **b** 7 – 1 = ☐ **c** 10 – 3 = ☐

d 5 – 4 = ☐ **e** 6 – 3 = ☐ **f** 9 – 8 = ☐

g 9 – 4 = ☐ **h** 9 – 7 = ☐ **i** 10 – 6 = ☐ **j** 7 – 3 = ☐ **k** 8 – 5 = ☐

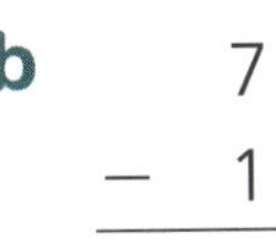

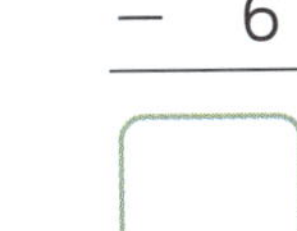

3

a 8 pegs – 4 pegs = ☐

b 8 bugs – 2 bugs = ☐

c 9 cats – 2 cats = ☐

d 10 bears – 5 bears = ☐

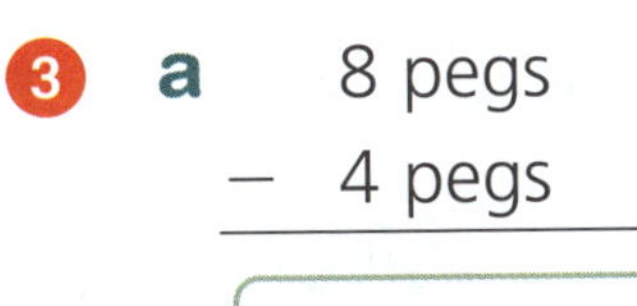
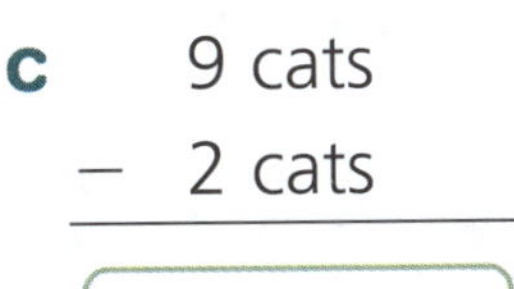
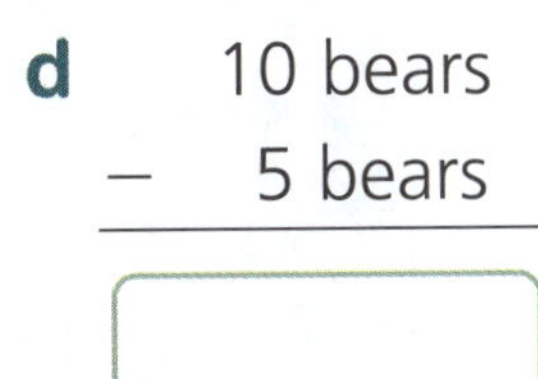

4

a

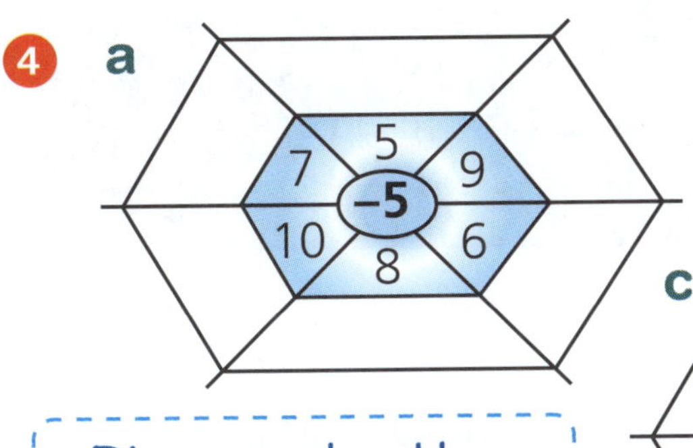

b

c

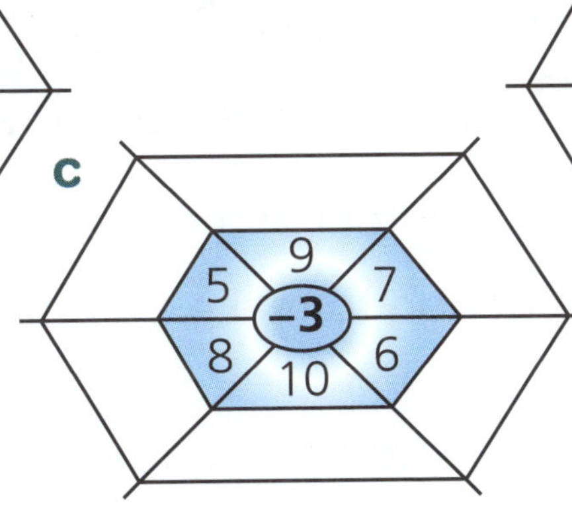

Discuss word problems for the number sentences on this page.

 • *AUSTRALIAN SIGNPOST MATHS NSW 2* • ISBN 9780655709039

1C Position words

beside	next to	left	right	centre	middle
above	on	onto	on top of	below	beneath
under	underneath	bottom	back to back	in front of	upside down
behind	between	close to	forward	near	far
further away	up	high	low	in	inside

1. Use some of these words to give the position of:

   a the emu

   b the bird

   c the rocks

   d the kangaroo lying down

   e the kangaroo standing up

   f the long grass plant

2. Describe the position of animals in this picture.

For example, the kangaroo (13) is in front of the tree.

3. **a** Use plasticine or playdough to make a simple model of the picture in Question 1.

   b Make a model of your bedroom or another room at home.

 • *AUSTRALIAN SIGNPOST MATHS NSW 2* • ISBN 9780655709039

1D Modelling numbers

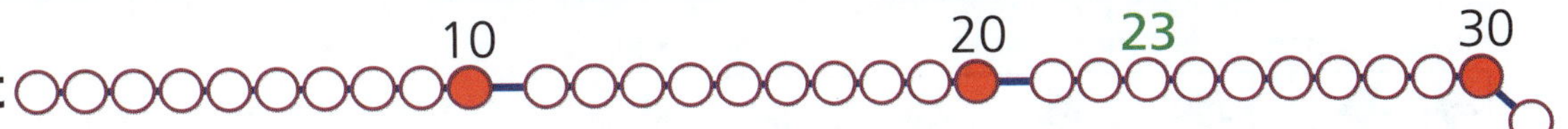

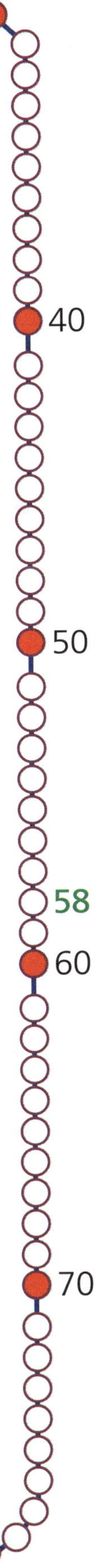

Start 10 20 23 30 40 50 58 60 70

The number 85 can be shown in different ways.

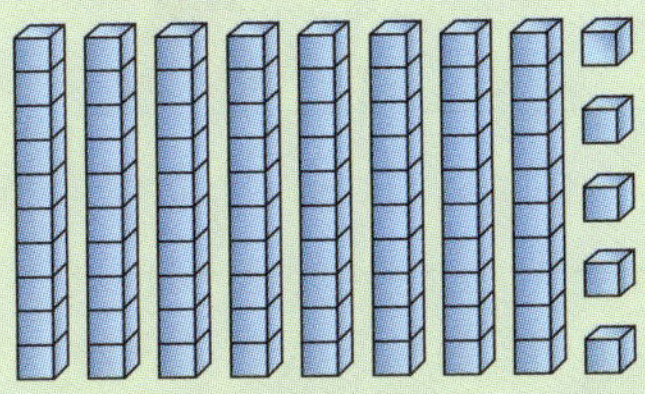

blocks

8 tens 5 ones

numeral expander

85 is also shown on the bead string at the bottom of the page.

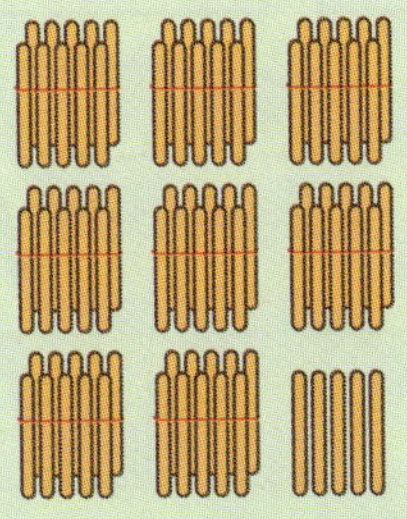

craft sticks

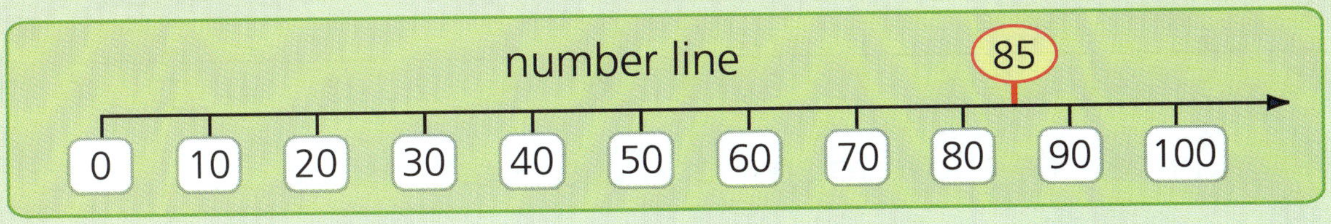

number line

85

0 10 20 30 40 50 60 70 80 90 100

1 Write the number modelled and fill in the numeral expander.

a

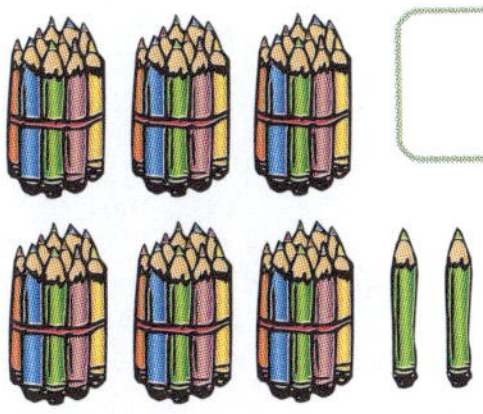

tens ones

b

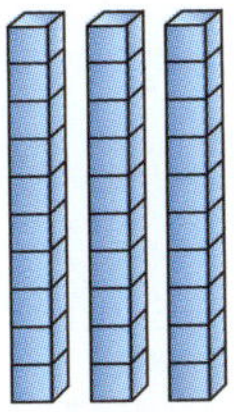

tens ones

c

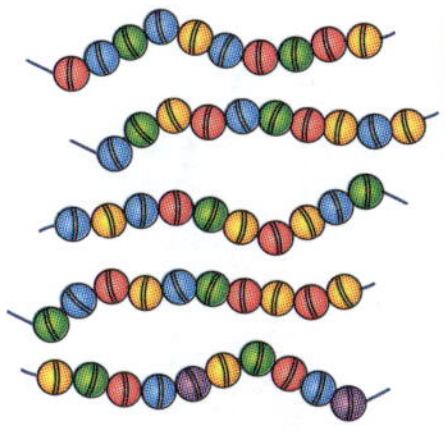

tens ones

d

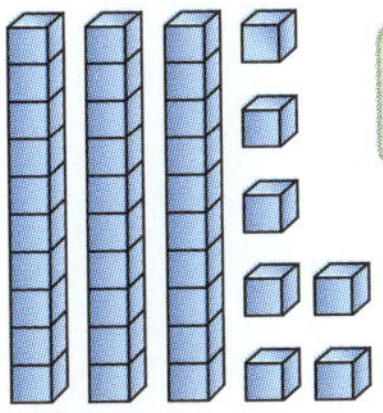

tens ones

e

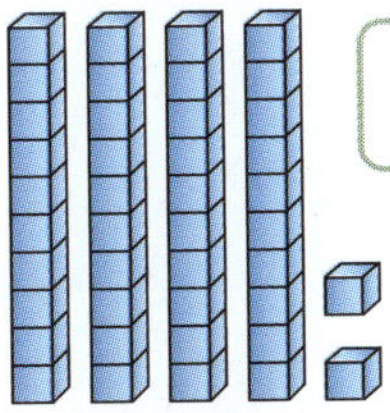

tens ones

f

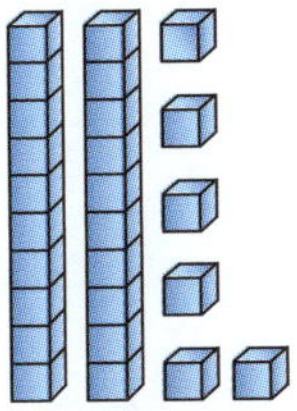

tens ones

2 Every 10th bead around this page is coloured red. Colour bead number:

a 16 b 35 c 49 d 62 e 74 f 92

3 Tick the bead that comes before:

a 10 b 30 c 65 d 78 e 95 f 108

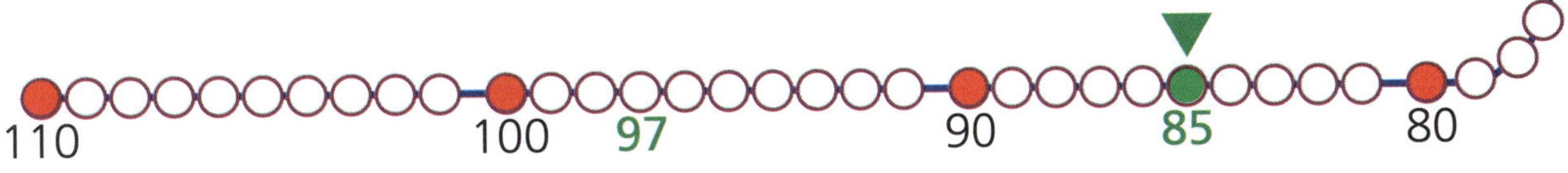

110 100 97 90 85 80

2A Addition

2 + 4 is the same as 4 + 2

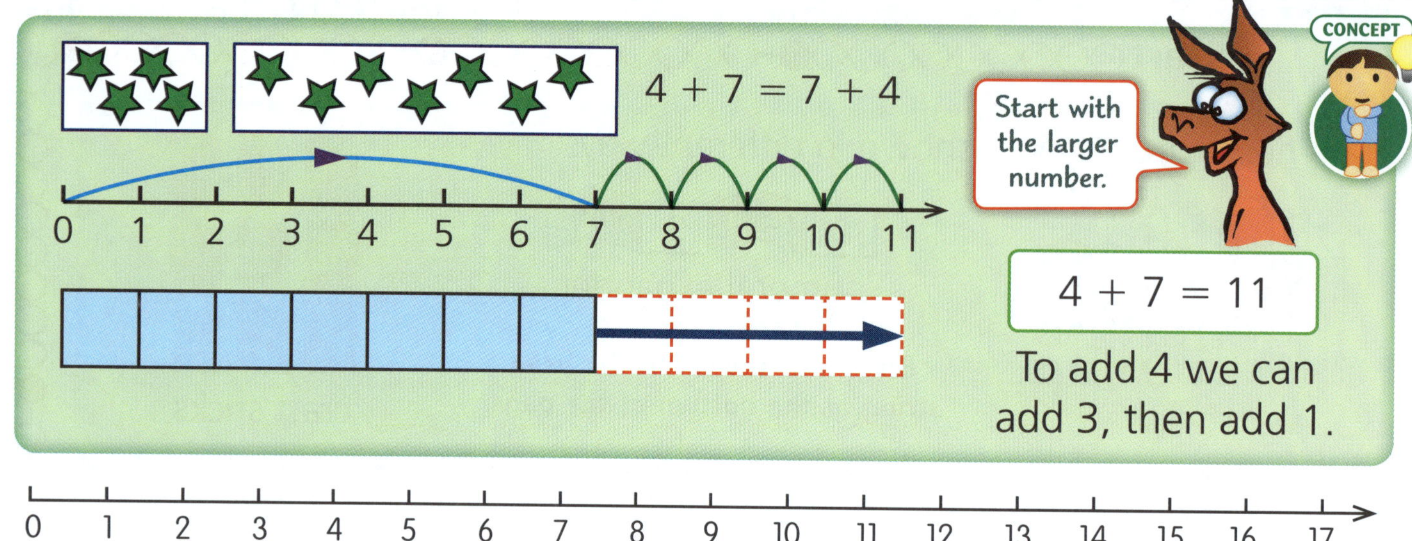

0 1 2 3 4 5 6 7 8 9 10 11 12 13 14 15 16 17

1 Use the number line above to answer these questions.

a 3 and 6 makes ☐.

b 5 and 5 makes ☐.

c $9 + 2 =$ ☐

d $7 + 4 =$ ☐

e $8 + 5 =$ ☐

f $8 + 3 =$ ☐

g $7 + 7 =$ ☐

h $9 + 4 =$ ☐

i $8 + 7 =$ ☐

j $7 + 8 =$ ☐

k $9 + 8 =$ ☐

l $8 + 6 =$ ☐

m $4 + 7 =$ ☐

n $2 + 9 =$ ☐

Use the number line to add to ten then add on the rest.

Example:
$7 + \mathbf{5} = (7 + \mathbf{3}) + \mathbf{2}$
$= 12$

0 1 2 3 4 5 6 7 8 9 10 11 12 13 (+3, +2)

o $6 + 7 =$ ☐

p $9 + 5 =$ ☐

q $5 + 7 =$ ☐

r $6 + 5 =$ ☐

s $4 + 8 =$ ☐

t $3 + 9 =$ ☐

Number and algebra

2B Addition to 20

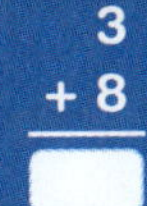

3 + 8 = ☐

1 Use counters or counting on to complete each number sentence.

a 4 + 9 = ☐ **b** 6 + 8 = ☐ **c** 7 + 9 = ☐

d 7 + ☐ = 14 **e** 9 + ☐ = 15 **f** 8 + ☐ = 12

g 5 + ☐ = 13 **h** 9 + ☐ = 18 **i** 10 + ☐ = 18

2

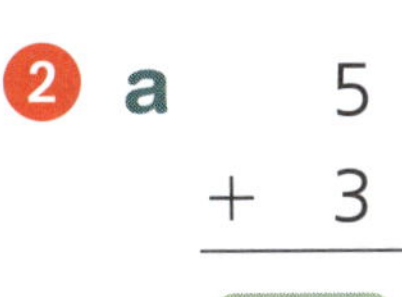

a 5 + 3 = ☐ **b** 8 + 4 = ☐ **c** 7 + 8 = ☐

d 5 + 9 = ☐ **e** 9 + 9 = ☐ **f** 7 + 6 = ☐

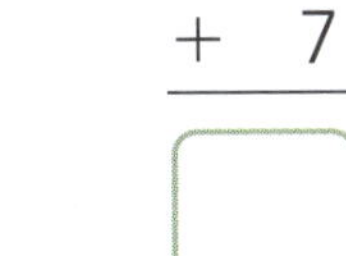

g 6 + 7 = ☐ **h** 10 + 10 = ☐ **i** 12 + 3 = ☐

We read this as 6 + 4 = 10

6
+ 4
10

Learn number facts to 20.

11	
8	☐
☐	7
6	☐
3	☐
2	☐
5	☐
☐	4

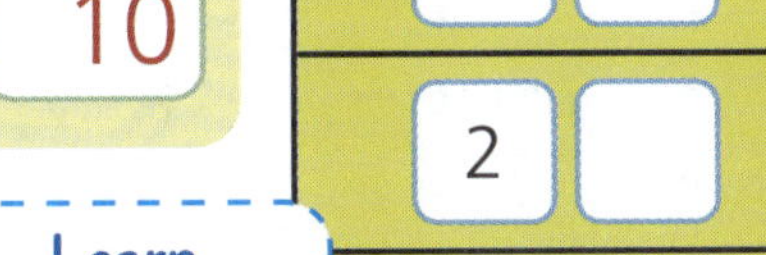

3 Write a different true number sentence each time.

a 12 = ☐ + ☐ **b** 15 = ☐ + ☐

12 = ☐ + ☐ 15 = ☐ + ☐

12 = ☐ + ☐ 15 = ☐ + ☐

12 = ☐ + ☐ 15 = ☐ + ☐

4 Complete these number bond houses.

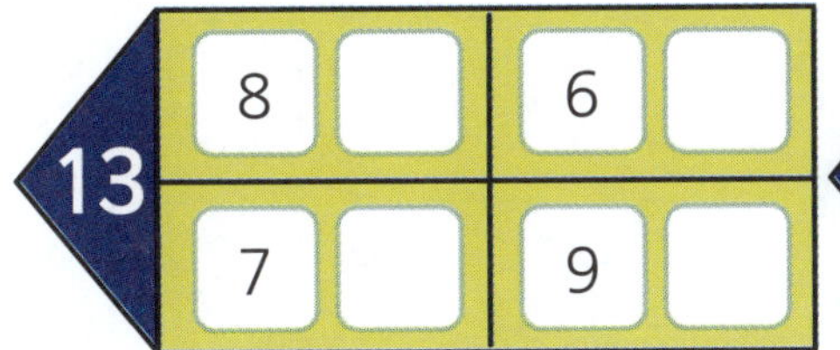

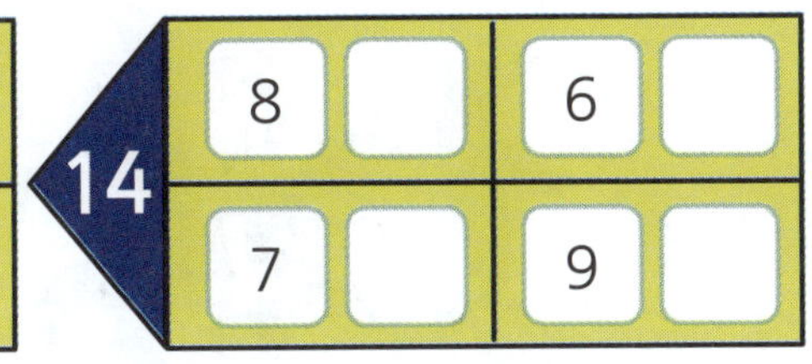

2C Addition to 20

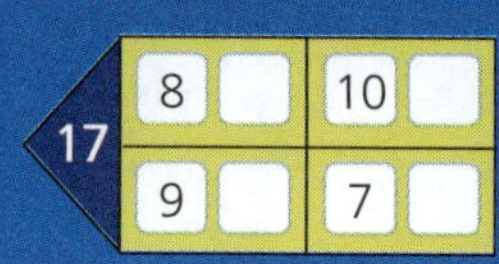
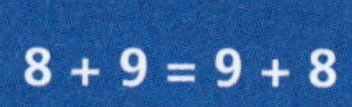

CONCEPT

8 + 5 Start at 8 and count on.

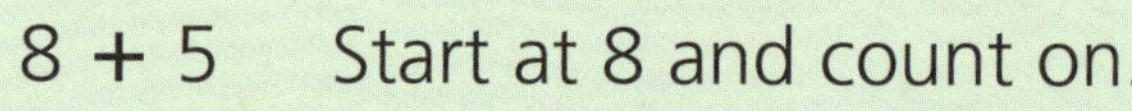
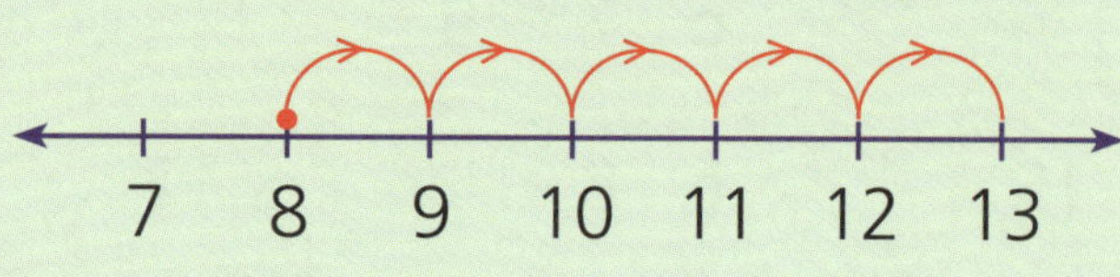

$$\begin{array}{r} 8 \\ +\ 5 \\ \hline 13 \end{array}$$

1 Use the number line to count on from the larger number.

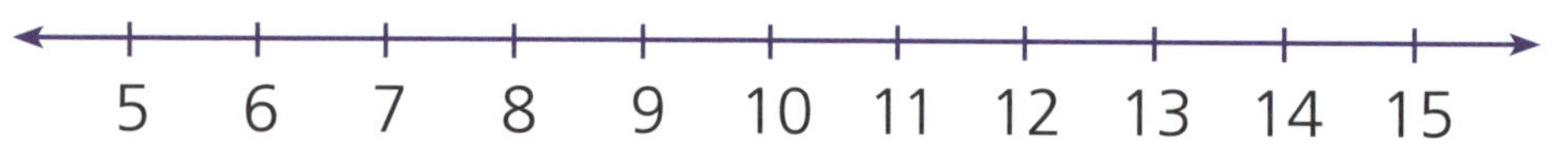

a 7 + 6 = ☐ **b** 4 + 7 = ☐ **c** 3 + 9 = ☐ **d** 8 + 4 = ☐ **e** 5 + 6 = ☐

f 4 + 8 = ☐ **g** 9 + 2 = ☐ **h** 6 + 8 = ☐ **i** 2 + 9 = ☐ **j** 7 + 8 = ☐

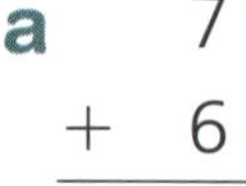

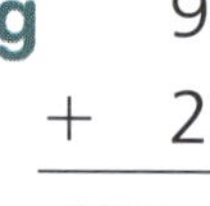

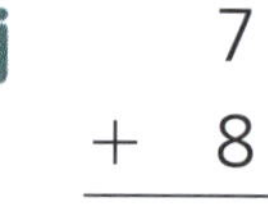

Learn your addition tables.

2 Use counters to find numbers for the windows.

3 **a** An ant has 6 legs.
Find different ways its legs can be grouped.

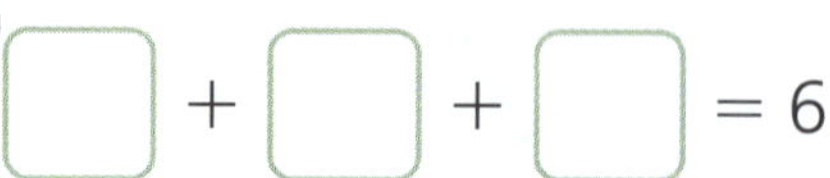

☐ + ☐ + ☐ = 6

☐ + ☐ + ☐ = 6

b An octopus has 8 legs.
Find different ways its legs can be grouped.

☐ + ☐ + ☐ = 8

☐ + ☐ + ☐ = 8

 • *AUSTRALIAN SIGNPOST MATHS NSW 2* • ISBN 9780655709039

2D Thinking about graphs

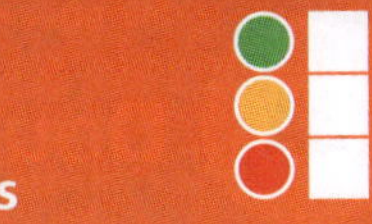

Words: baseline
equal spacing
same-sized symbols

1

A

B

In the pond

Tadpoles

Frogs

Fish

Which of these do you think is the better picture graph?

Why do you think so?

2 **a** Each smiley face stands for a sticker.
Discuss what is wrong with this graph.

Number of stickers given

Alana Deepak Brand

b Complete this table.

Name	Number
Alana	
Deepak	
Brand	

c Draw a better graph.
Colour one face for each sticker.

Number of stickers given

Alana Deepak Brand

3A Doubling and near doubling

6 + 6
7 + 7
8 + 8
9 + 9

CONCEPT

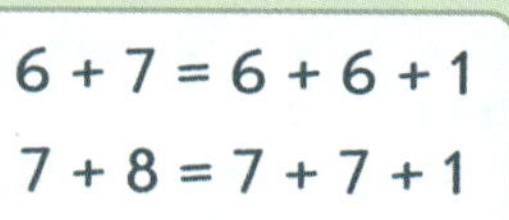

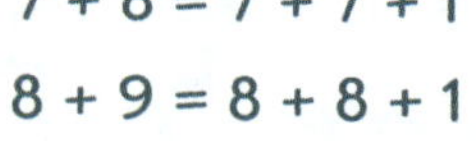

$6 + 7 = 6 + 6 + 1$

$7 + 8 = 7 + 7 + 1$

$8 + 9 = 8 + 8 + 1$

When you **double** a number, the answer will be **even**.

1 Complete each number sentence.

a double 3
☐ + ☐ = ☐

b double 6
☐ + ☐ = ☐

c double 1
☐ + ☐ = ☐

d double 7
☐ + ☐ = ☐

e double 9
☐ + ☐ = ☐

f double 8
☐ + ☐ = ☐

2

a $2 + 2 = $ ☐

b $4 + 4 = $ ☐

c $0 + 0 = $ ☐

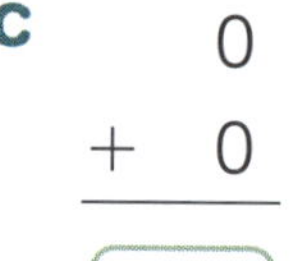

d $5 + 5 = $ ☐

e $10 + 10 = $ ☐

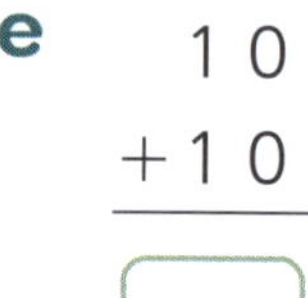

3

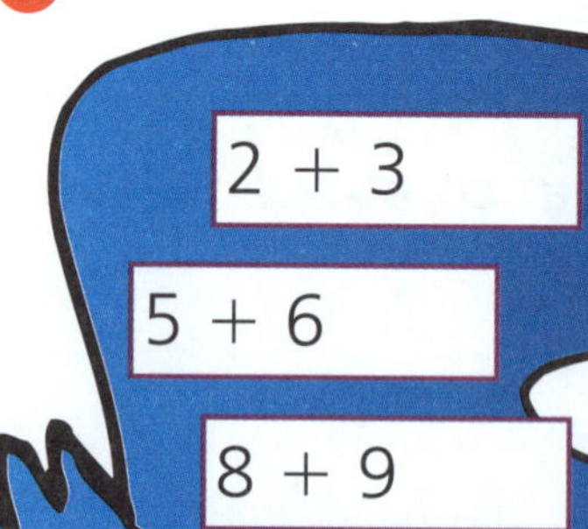

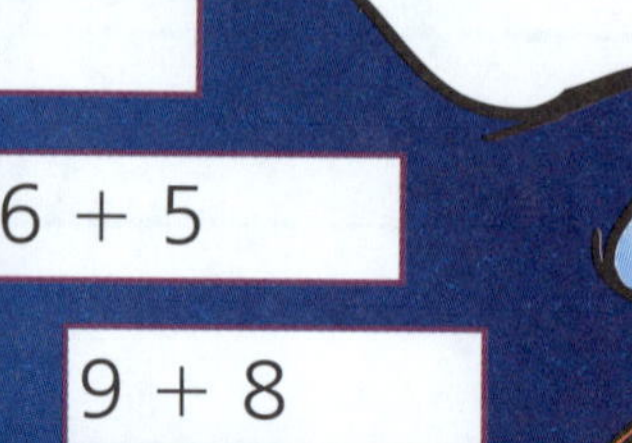

2 + 3
5 + 6
8 + 9
3 + 4
3 + 2
4 + 5
6 + 5
6 + 7
9 + 8
7 + 8

 • *AUSTRALIAN SIGNPOST MATHS NSW 2* • ISBN 9780655709039

3B Sharing

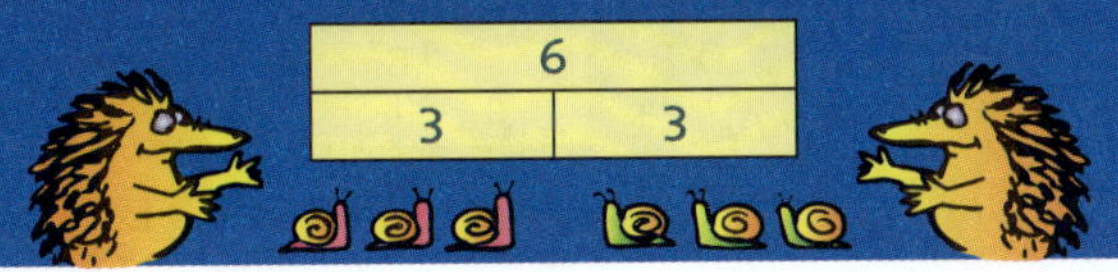

CONCEPT

When sharing fairly, each person is given the same amount.
Some counters could be left over.

?	?	?
9		

We all have a fair share.

1. Share 6 counters among 3 boxes. One share = ☐

   ☐ ☐ ☐

   If we combine the shares again, how many counters would we have? ☐

2. Share 20 counters between 2 groups.

   ☐ ☐ One share = ☐

   If we combine the shares again, how many counters would we have? ☐

3. Share 25 counters between 2 groups.

   ☐ ☐ One share = ☐

   and ☐ left over

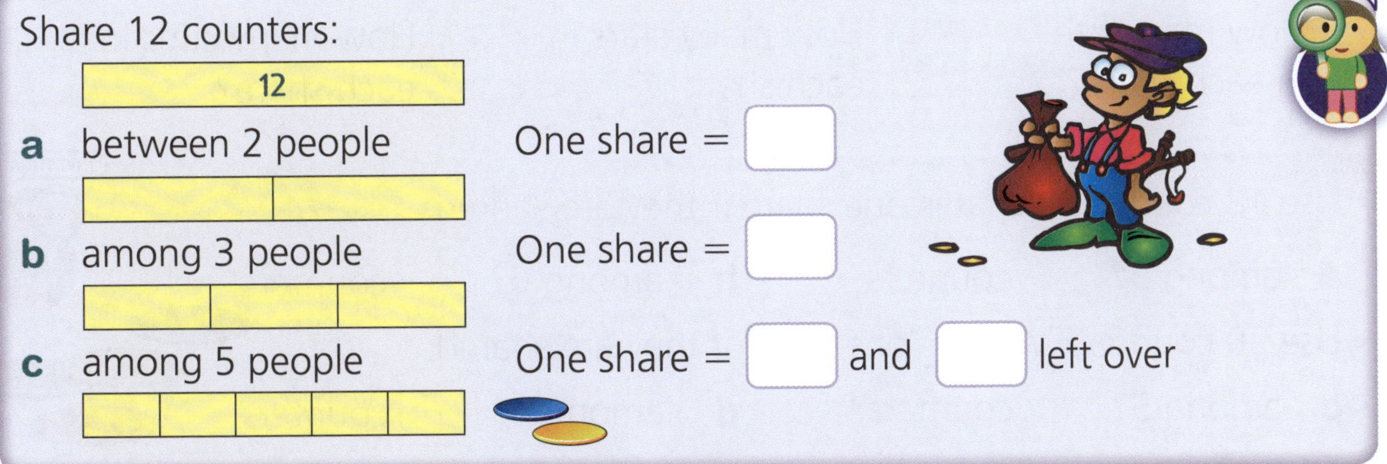

INVESTIGATION

Share 12 counters:

12

a between 2 people One share = ☐

b among 3 people One share = ☐

c among 5 people One share = ☐ and ☐ left over

 • *AUSTRALIAN SIGNPOST MATHS NSW 2* • ISBN 9780655709039

Sharing

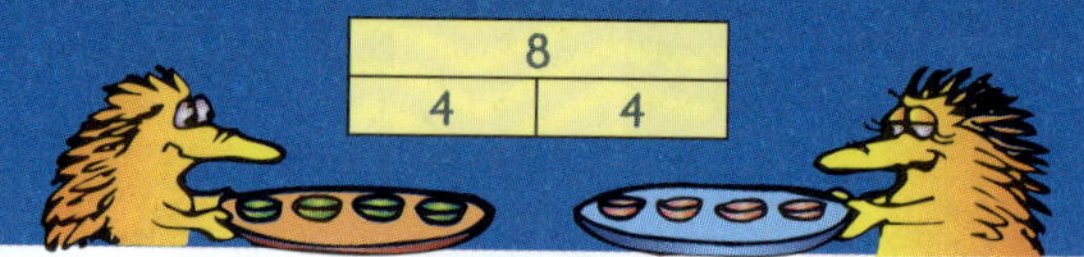

When we share, each person is given the same number of items.

1 Circle each share, then answer the questions.

a Shared among 4

How many stars?

How many shares?

How many stars in each share?

b Shared among 4

How many caps?

How many shares?

How many caps in each share?

c Shared between 2

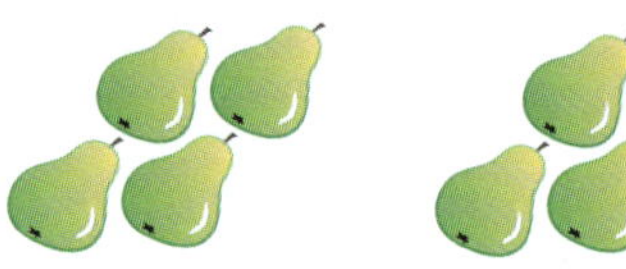

How many pears?

How many shares?

How many pears in each share?

d Shared among 4

How many fish?

How many shares?

How many fish in each share?

e Shared between 2

How many stars?

How many shares?

How many stars in each share?

f Shared among 4

How many cups?

How many shares?

How many cups in each share?

Use 18 counters. What is one share if they are shared:

a among 3? ☐ counters **b** among 6? ☐ counters

Use 20 counters. What is one share if they are shared:

c among 5? ☐ counters **d** among 4? ☐ counters

3D 2D shapes

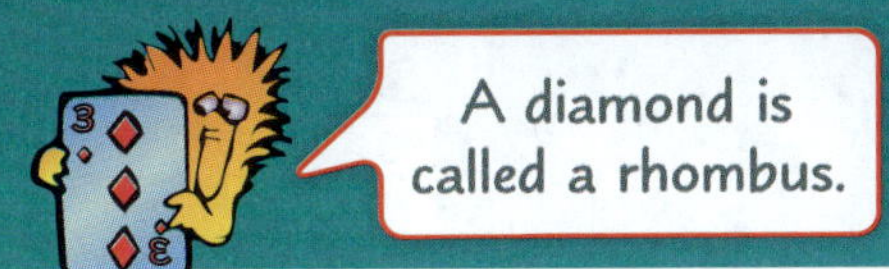

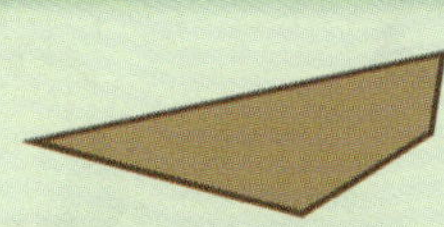

Quadrilateral: 4 straight sides
4 vertices (corners)

Rectangle: opposite sides equal
4 vertices (corners)

Hexagon: 6 straight sides
6 vertices (corners)

1. In this picture, find how many:

   a circles ☐

   b triangles ☐

   c squares ☐

   d hexagons ☐

   e pentagons ☐

   f octagons ☐

   Discuss their features.

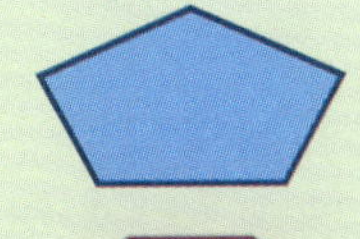

Pentagon: 5 straight sides, 5 vertices

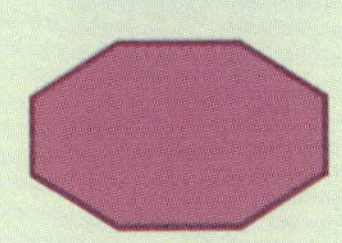

Octagon: 8 straight sides, 8 vertices

2. Write the name of each shape.

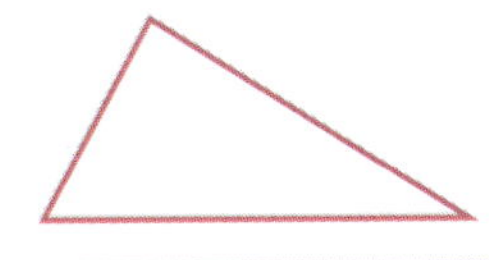

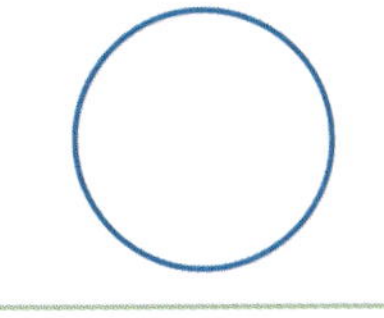

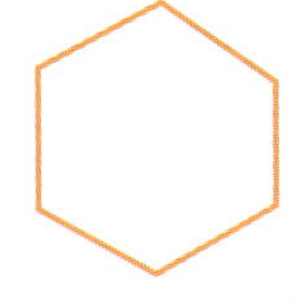

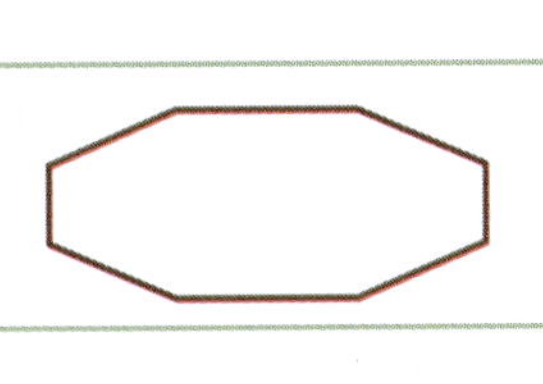

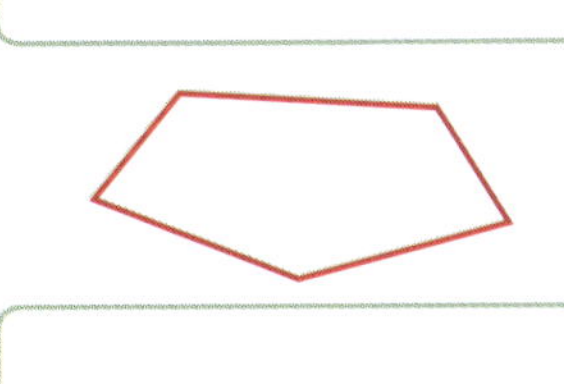

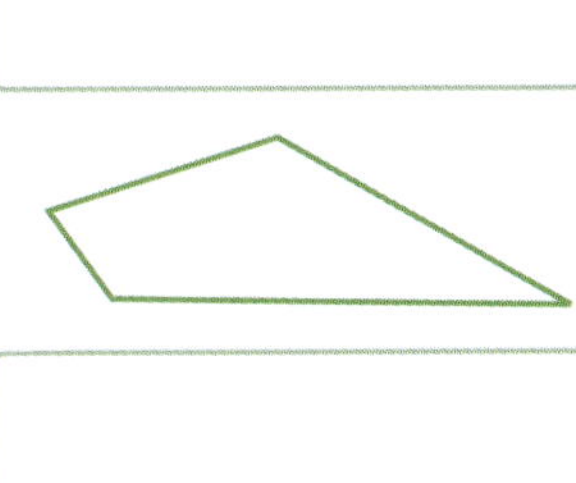

Subtraction

If 6 + 8 = 14 then 14 – 8 = 6.

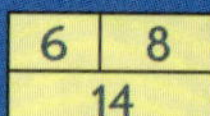

CONCEPT

11 take away 4 leaves 7.

11 – 4 = 7

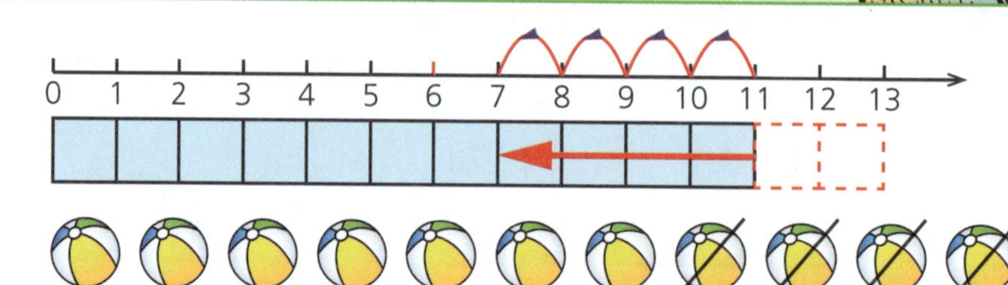

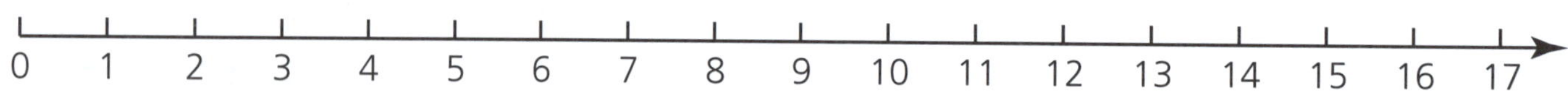

1 Use the number line above to answer these questions.

a 13 minus 4 equals ☐.
b 12 minus 3 equals ☐.

c 11 – 6 = ☐
d 15 – 8 = ☐
e 12 – 4 = ☐

f 10 – 7 = ☐
g 14 – 7 = ☐
h 16 – 9 = ☐

i 11 – 3 = ☐
j 13 – 6 = ☐
k 13 – 5 = ☐

l 17 – 9 = ☐
m 16 – 8 = ☐
n 14 – 6 = ☐

You can subtract to the 10, then take away the rest.

13 – **7** = (13 – **3**) – **4**
= 6

–4 –3
0 1 2 3 4 5 6 7 8 9 10 11 12 13

o 12 – 7 = ☐
p 13 – 8 = ☐
q 17 – 8 = ☐

r 14 – 8 = ☐
s 16 – 7 = ☐
t 15 – 6 = ☐

Subtraction to 20

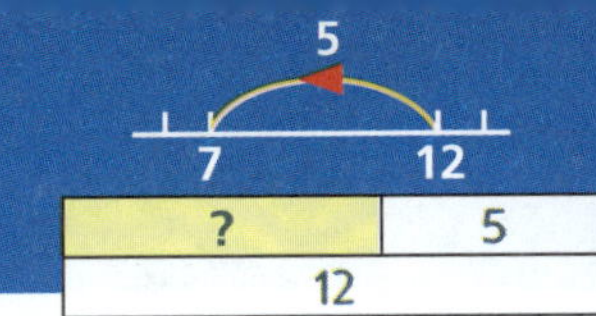

CONCEPT

Billy had 12 apples.

He ate 5.

How many did he have left?

12 – 5 = 7

$$\begin{array}{r} 12 \\ -\ 5 \\ \hline 7 \end{array}$$

1 **a** 10 bottles

8 broken

How many left?

☐ – ☐ = ☐

b 12 flowers

7 picked

How many left?

☐ – ☐ = ☐

c 9 cakes

6 eaten

How many left?

☐ – ☐ = ☐

d 15 balls

6 lost

How many left?

e 13 pencils

8 broken

How many left?

f 14 mice

7 caught

How many left?

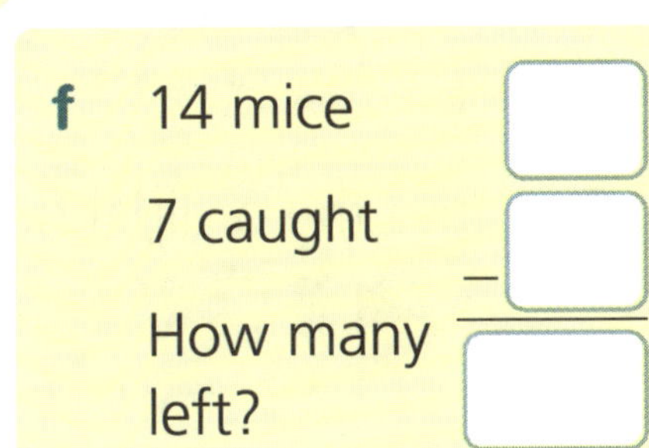

g 16 nuts

9 eaten

How many left?

Ask: What plus 9 makes 16?

FUN SPOT

Challenge a friend to answer these questions quickly.

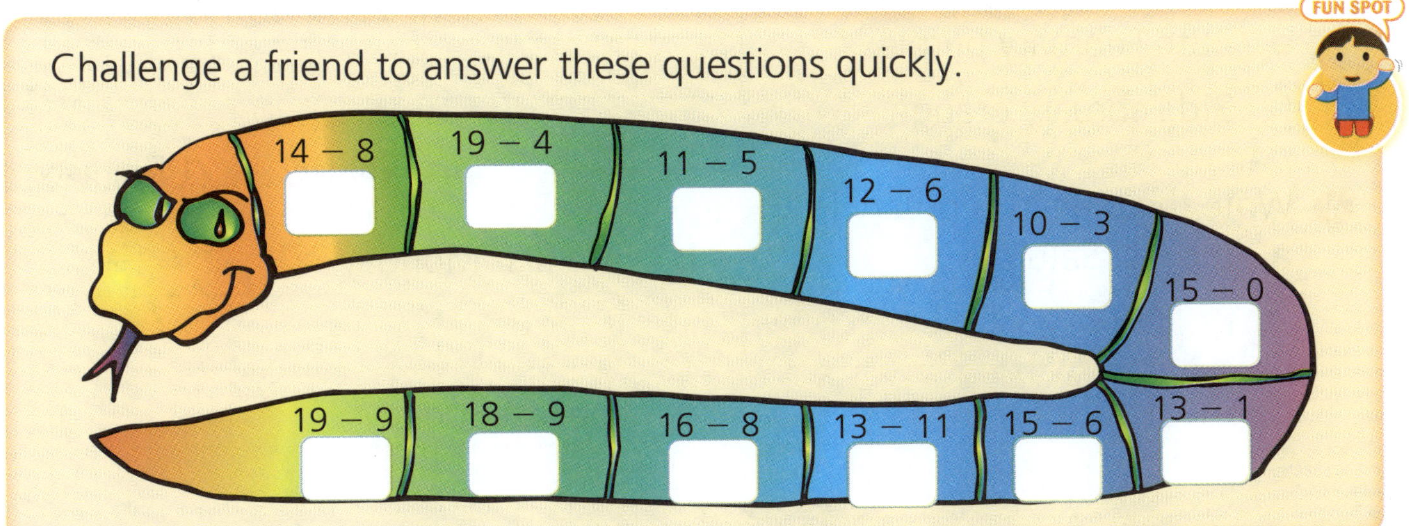

 • *AUSTRALIAN SIGNPOST MATHS NSW 2* • ISBN 9780655709039

4C Ordinal numbers and calendars

1 a Write the ordinal number for each dinosaur.

1st

Colour:

b the 3rd dinosaur green

c the 6th dinosaur red

d the 2nd dinosaur blue

e the 10th dinosaur purple

f the 4th dinosaur orange

g the 1st dinosaur pink

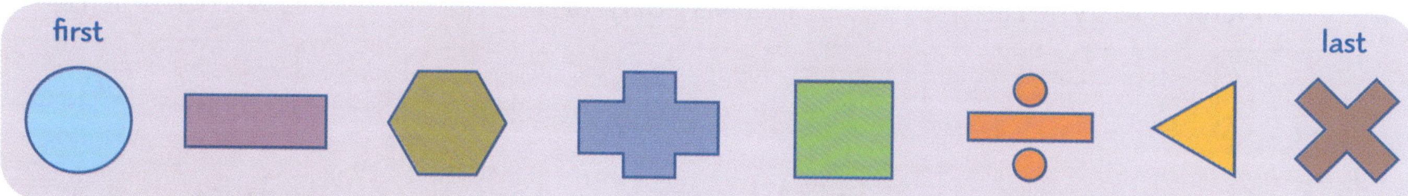

2 Write the ordinal number for:

a [triangle] b [square] c [hexagon] d [circle]

3 Colour:

a 21st February red

b 2nd February green

c 19th February blue

d 26th February yellow

e 28th February purple

f 3rd February orange

February						
Sun	Mon	Tues	Wed	Thurs	Fri	Sat
			1	2	3	4
5	6	7	8	9	10	11
12	13	14	15	16	17	18
19	20	21	22	23	24	25
26	27	28				

21st February = 21 February

4 Write the date in February for:

a the last Saturday

b the first Monday

c the first Friday

d the last Sunday

4D The calendar

Here is a calendar for October. Fill in the missing numbers.

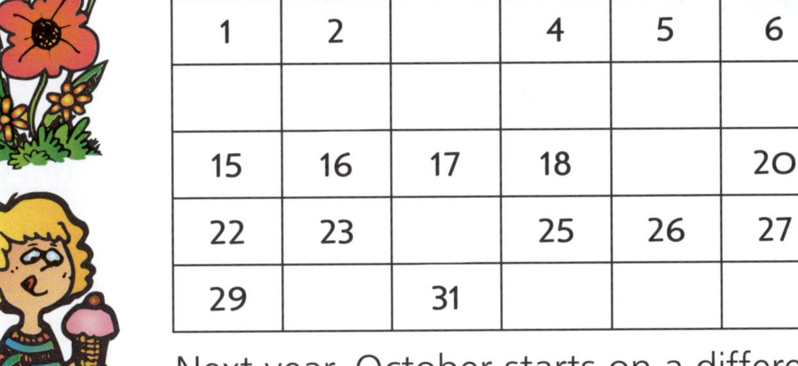

October						
Sun	Mon	Tues	Wed	Thurs	Fri	Sat
1	2		4	5	6	7
15	16	17	18		20	21
22	23		25	26	27	28
29		31				

30 days has September,
April, June and November.
All the rest have 31
Except February alone,
Which has 28 days clear,
And 29 days each leap year.

Next year, October starts on a different day.

1 a How many days are in October? ☐

b How many Sundays are there in this month? ☐

How many weeks is it from 2nd of October until:

c 23rd October? ☐ d 16th October? ☐ e 30th October? ☐

How many days is it from 4th of October until:

f 28th October? ☐ g 17th October? ☐ h 22nd October? ☐

On what day is:

i 4th October? ☐ j 21st October? ☐ k 29th October? ☐

2 Would you use minutes, hours, days or months to measure:

a the time it takes for a trip to the zoo? ☐

b the time it takes for eggs to hatch? ☐

c the time it takes to build a house? ☐

d the time it takes to cook a sausage? ☐

 • *AUSTRALIAN SIGNPOST MATHS NSW 2* • ISBN 9780655709039

5A Addition to 20

	+4	+5	+6	+7	+8	+9
8	12	13	14	15	16	17
9	13	14	15	16	17	18

1 Draw a picture and complete the number sentences.

a Milly got 8 pencils from Dad, 6 from Mum and 5 from Bess. How many did Milly get altogether?

☐ + ☐ + ☐ = ☐

b Ben had 6 DVDs. Dom gave him 4 DVDs and Ravi gave him 8. How many DVDs does Ben have altogether?

☐ + ☐ + ☐ = ☐

2 Count on or use doubles to complete.

a 3 + 4 + 3 = ☐ **b** 10 + 2 + 2 = ☐ **c** 6 + 1 + 6 = ☐

d 4 + 3 + 5 = ☐ **e** 8 + 4 + 2 = ☐ **f** 6 + 4 + 4 = ☐

g 4 + 3 + 4 = ☐ **h** 1 + 6 + 4 = ☐ **i** 3 + 6 + 2 = ☐

3 Write three number sentences for each total.

a
☐ + ☐ =
☐ + ☐ = **15**
☐ + ☐ =

b
☐ + ☐ =
☐ + ☐ = **17**
☐ + ☐ =

c
☐ + ☐ =
☐ + ☐ = **16**
☐ + ☐ =

d
☐ + ☐ =
☐ + ☐ = **18**
☐ + ☐ =

4 Complete these addition facts.

	+4	+5	+6	+7	+8	+9
6						
7						

	+8	+9
8		
9		

5B Addition by looking for tens

Do you know the friends of ten?

1

a 6, 4, 6, + 4	**b** 3, 3, 2, + 7	**c** 5, 9, 3, + 1			
d 7, 3, 4, + 5	**e** 4, 8, 2, + 2	**f** 9, 7, 1, + 2			
g 4, 5, 6, + 7	**h** 3, 9, 8, + 1	**i** 5, 7, 8, + 5	**j** 3, 6, 9, + 4	**k** 6, 6, 5, + 4	

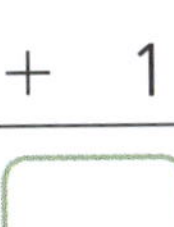

2

a \$5, \$7, \$5, + \$3	**b** \$9, \$3, \$1, + \$2	**c** \$2, \$6, \$7, + \$3	**d** \$8, \$4, \$2, + \$7	**e** \$7, \$6, \$4, + \$5

3 **a** 8 cows, 3 sheep, 2 goats, 7 pigs. How many animals? ☐

b 2 fish, 9 mice, 7 birds, 1 emu. What is the total number? ☐

4 True (T) or false (F)?

a $6 + 7 + 3 + 4 + 5 = 5 + 4 + 7 + 6 + 3$ ☐

b $2 + 5 + 8 + 5 + 3 = 3 + 8 + 2 + 3 + 7$ ☐

c $3 + 9 + 1 + 9 + 7 = 9 + 5 + 2 + 8 + 5$ ☐

 • *AUSTRALIAN SIGNPOST MATHS NSW 2* • ISBN 9780655709039

5C Looking at 3D objects

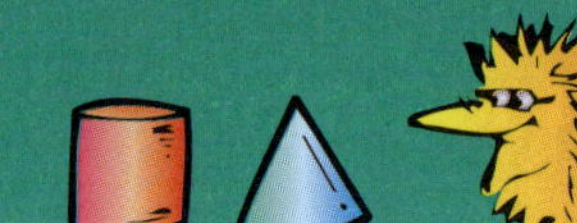

1. Write some examples of spheres.

2. Choose words from the list that describe each object.

straight
rounded
smooth
flat
pointed
curved

ACTIVITY

- Draw a sphere.
- Draw a 3D object that has a point.
- Make a sphere using plasticine or playdough.

 • *AUSTRALIAN SIGNPOST MATHS NSW 2* • ISBN 9780655709039

5D Describing 3D objects

1 Complete:

	Slides	Rolls	Stacks	Sinks	Floats
Balloon	✗	✓	✗	✗	✓
Nut					
Cork					
Block					
Can					
Marble					

2 Colour the objects that match the description.

6 square faces	Fitwell Shoes
No corners (vertices)	Juice; FIZZ ORANGE

INVESTIGATION

Find objects in the classroom to complete the table.

Object	Curved surface(s)	Flat surface(s)
	✓	
		✓
	✓	✓

Sharing and grouping

For grouping, we ask how many groups can be made?

CONCEPT

Sharing

8 stickers were shared between Zoe and Lakshmi.

They got **4** each.

Grouping

I had 8 stickers. How many people could take 2 stickers?

4 groups of 2 makes 8.

4 people could be given stickers.

1 Draw lines to show how you would share these collections equally.

a 6 apples
2 children

☐ each

b 10 stickers
2 children

☐ each

c 12 balls
3 children

☐ each

2 Draw lines to show the groups that have been made.

a

6 apples.
How many children could take 2 apples?

b

10 stickers.
How many children could take 2 stickers?

c

12 balls.
How many children could take 3 balls?

3 Show how you could divide this array into two equal groups. Write what you did.

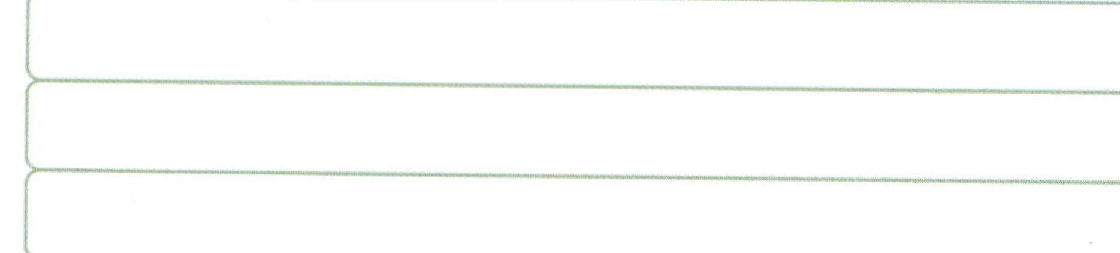

 • *AUSTRALIAN SIGNPOST MATHS NSW 2* • ISBN 9780655709039

6B Groups and rows

You can skip jump.

1 Use a number line or number chart to complete these.

a

☐ + ☐ + ☐

3 groups of 3 = ☐

b

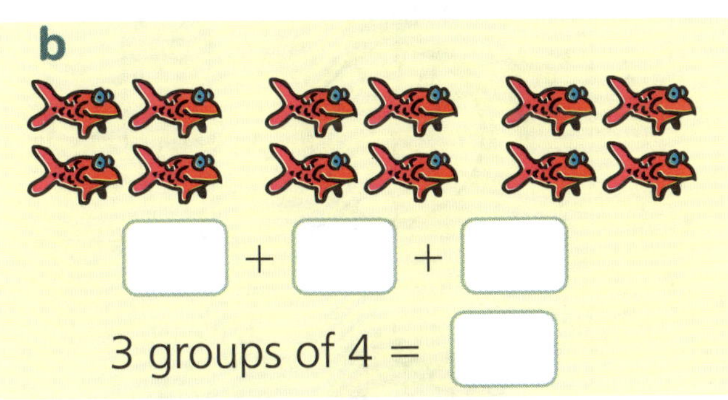

☐ + ☐ + ☐

3 groups of 4 = ☐

c

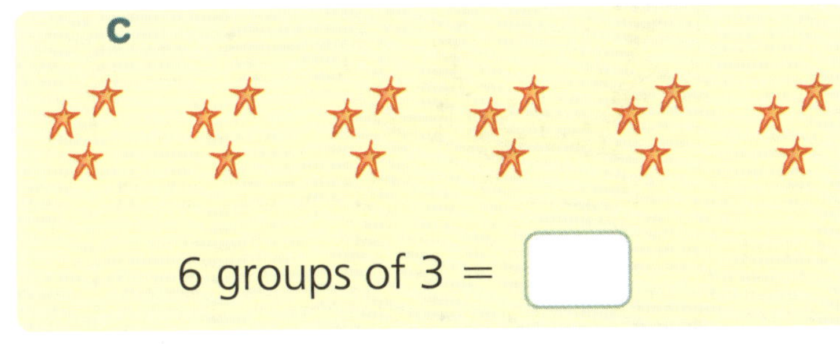

6 groups of 3 = ☐

d

4 rows of 5 = ☐ or

5 columns of 4 = ☐

e

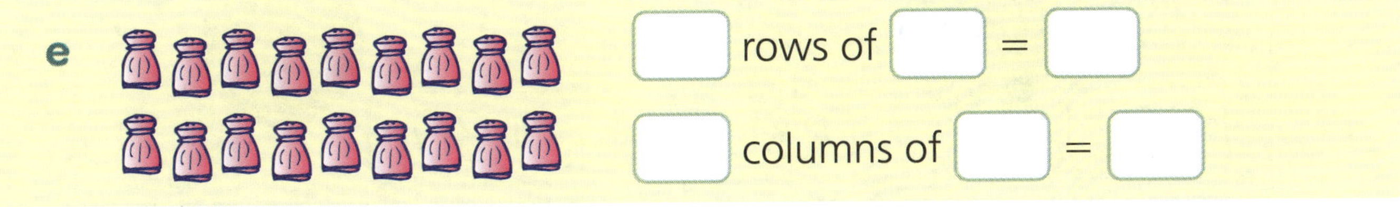

☐ rows of ☐ = ☐

☐ columns of ☐ = ☐

2 Use counters to solve these problems.

a 3 groups of 5 balls = ☐ balls

b 2 rows of 7 trees = ☐ trees

c 3 groups of 8 ants = ☐ ants

d 4 rows of 6 people = ☐ people

 • *AUSTRALIAN SIGNPOST MATHS NSW 2* • ISBN 9780655709039

6C Revision of time

half past 8
8 thirty

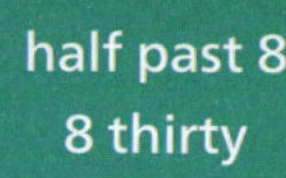

CONCEPT

Both clocks show 2 o'clock.

Both clocks show half past 12.

1 Read the time and complete the sentence. Write the time on the digital clock.

The minute hand is longer.

The hour hand is shorter.

a

b

c

d

2 Match the analog time to the correct digital time.

a

b

FUN SPOT

- Show these times on the digital clock.

School starts : Recess :

Lunch : Home time :

- Find out how many seconds there are in one minute.

6D Estimating time passed

1 Work in groups of three. As one person writes their name, the other two clap. Record the number of claps.

Number of claps for friend 1	
Number of claps for friend 2	

Who took the least time to write their name?

2 Use clapping to time other events. First estimate the time it will take.

Event	Claps

3 Practise saying "1 second, 2 seconds, 3 seconds …" so that each count takes 1 second.

a Count like this for 1 minute (60 seconds). What number did you reach?

b Repeat the activity in Question 1 using this form of counting.

Number of seconds for friend 1	
Number of seconds for friend 2	

1 second, 2 seconds,
3 seconds, 4 seconds ...

4 Use counting to time other events. Estimate first.

	Event	Seconds
A		
B		

 • *AUSTRALIAN SIGNPOST MATHS NSW 2* • ISBN 9780655709039

7A Groups and rows

three groups of four
4 + 4 + 4

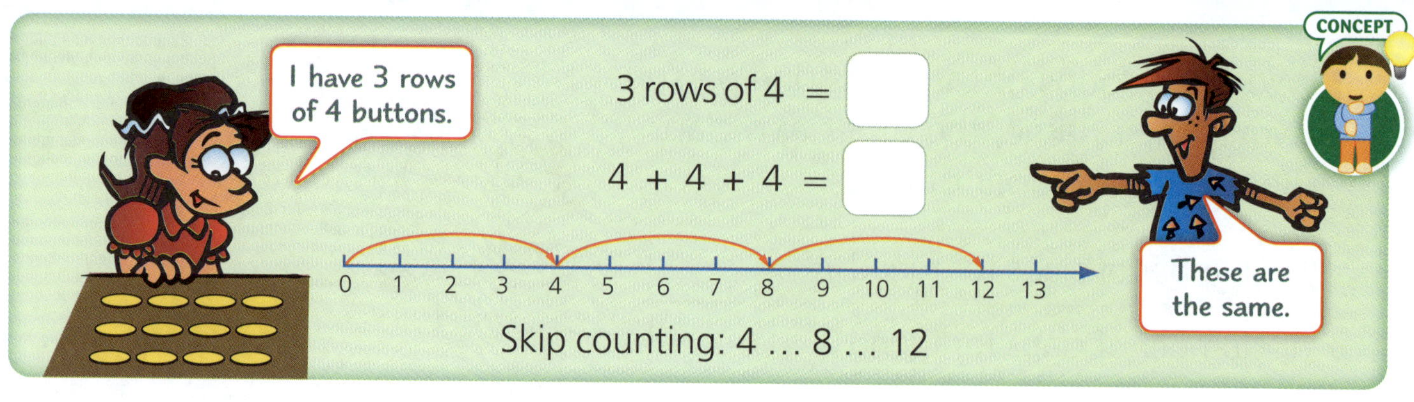

1 Complete:

a

2 groups of 3 = ☐

3 + 3 = ☐

b

☐ groups of ☐ = ☐

☐ + ☐ + ☐ = ☐

c

☐ rows of ☐ = ☐

☐ + ☐ + ☐ = ☐

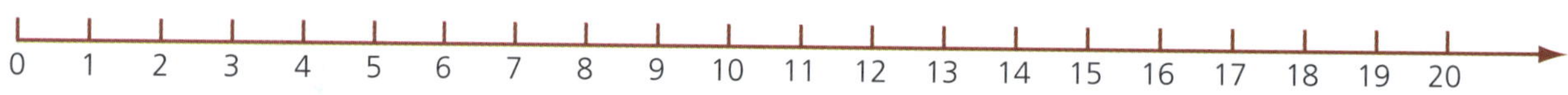

d

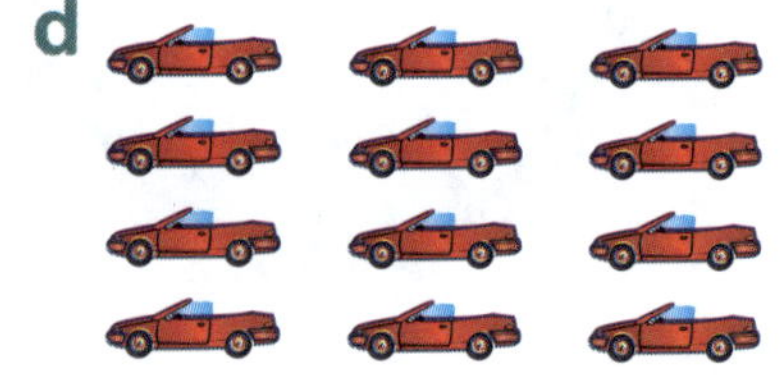

☐ rows of ☐ = ☐

e

☐ groups of ☐ = ☐

f

1 group of ☐ = ☐

g

☐ groups of ☐ = ☐

h

☐ groups of ☐ = ☐

i

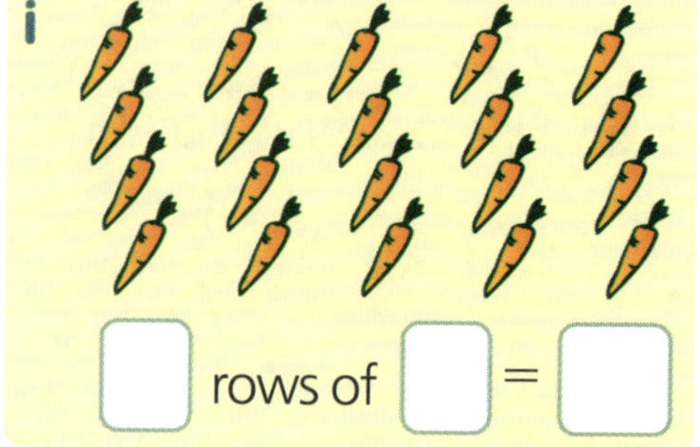

☐ rows of ☐ = ☐

7B Problem solving

1 Use the drawings above to complete these number sentences.

a ☐ groups of ☐ flying birds = ☐ flying birds

b ☐ group of ☐ cows = ☐ cows

c ☐ rows of ☐ frogs = ☐ frogs

d 1 group of ☐ cockatoos = ☐ cockatoos

e 1 group of ☐ fish = ☐ fish

f ☐ groups of ☐ turtles = ☐ turtles

g ☐ rows of ☐ trees = ☐ trees

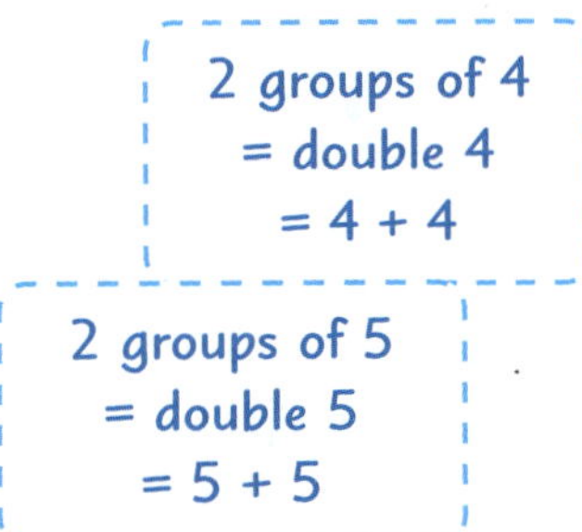

Use the groups above to make up other problems.

a 2 groups of 4 cockatoos = ☐ cockatoos

b ☐ groups of ☐ ☐ = ☐ ☐

c ☐ groups of ☐ ☐ = ☐ ☐

7C Features of 2D shapes

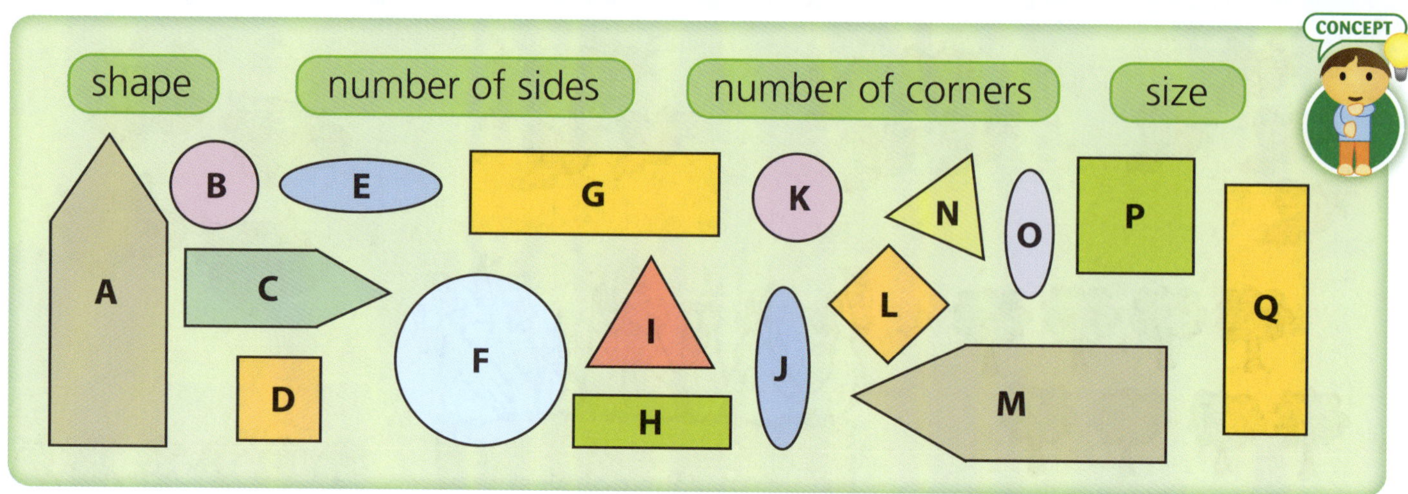

1 Look at these blocks.

a Which have the same shape as E?

b Which have the same shape as K?

c Which has the same shape as I?

d Which have the same shape as G?

e Which have the same shape and size as B?

f Which have the same shape and size as A?

g Which has the same shape and size as G?

h Which have the same shape as P?

i Which have the same shape as M?

j Which have four corners (vertices)?

2 What is the name of:

a shape A?

b shape B?

c shape Q?

d shape N?

e shape D?

f shape J?

 • *AUSTRALIAN SIGNPOST MATHS NSW 2* • ISBN 9780655709039

7D Drawing 2D shapes

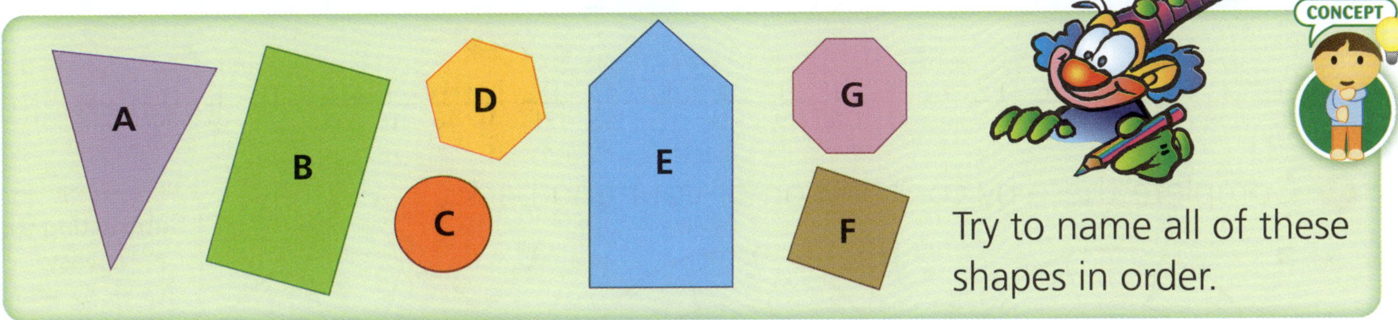

Try to name all of these shapes in order.

1 Use a ruler to draw these shapes. Use the dots to help.

a triangle **b** rectangle **c** hexagon

d pentagon **e** octagon **f** quadrilateral

ACTIVITY

Use the shapes to add to this design. Try to name shapes A to F.

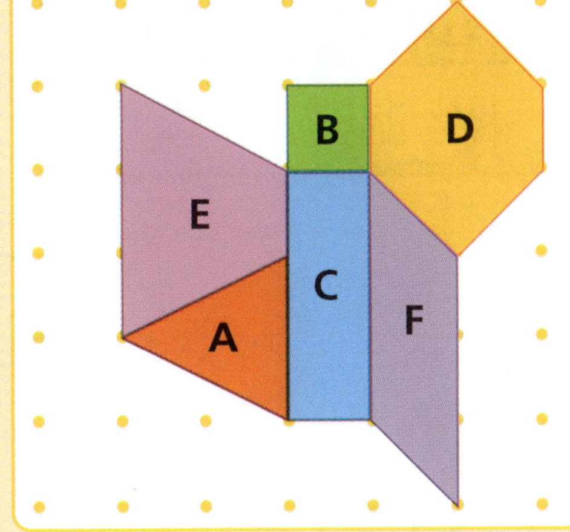

Subtraction to 20

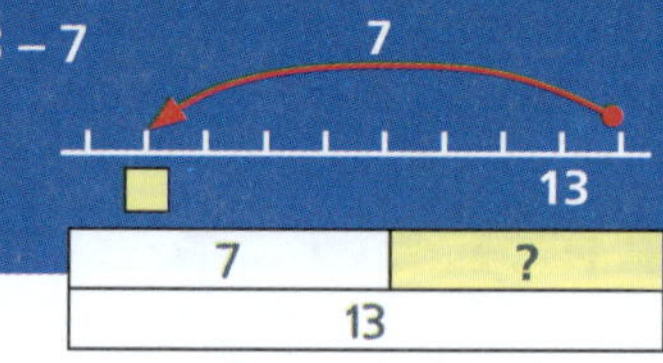

0 1 2 3 4 5 6 7 8 9 10 11 12 13 14 15 16 17 18 19 20

1 Complete these by counting on or counting back.

a

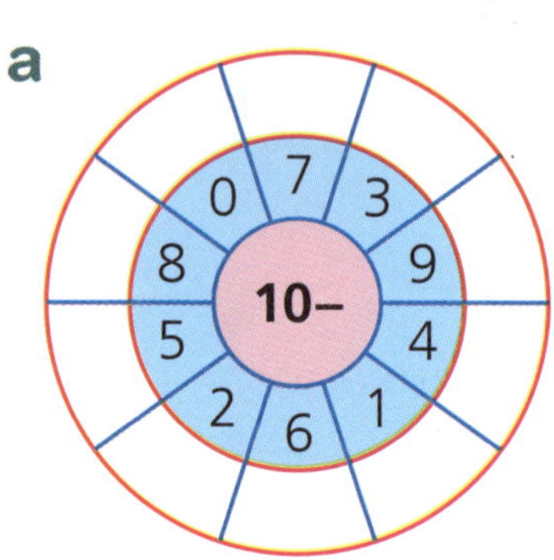

b

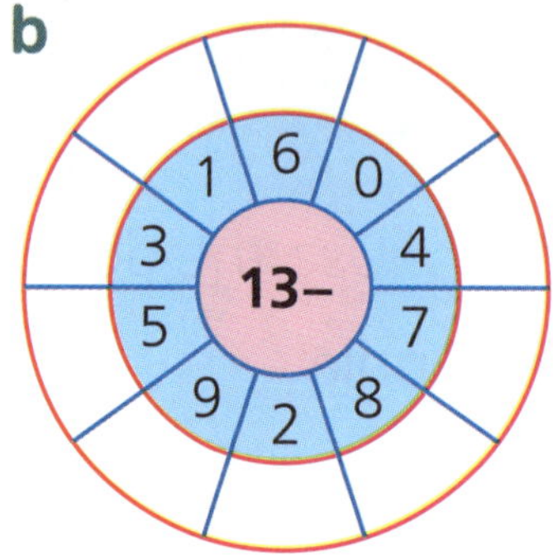

c

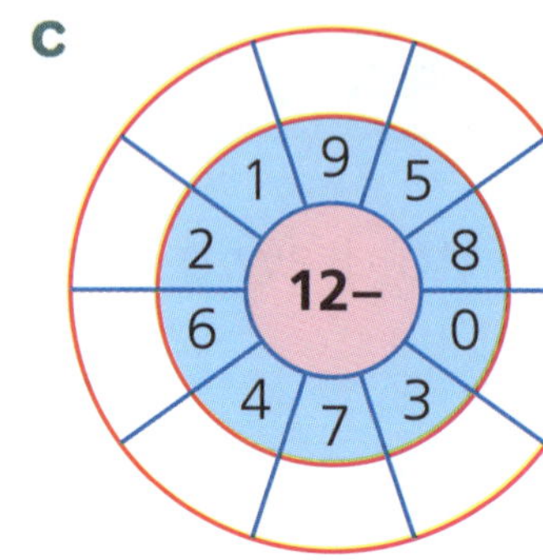

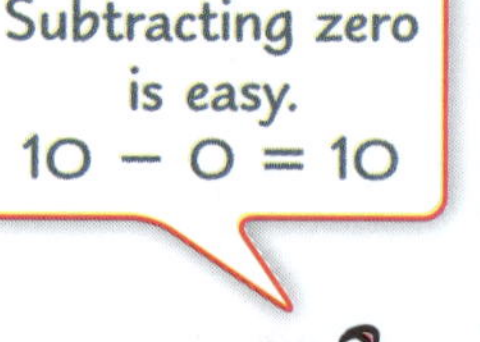

2 Colour the boxes. 10 = red, 9 = blue, 8 = green, 7 = yellow

14 – 4	13 – 4	9 – 2	13 – 6	11 – 4	16 – 7	20 – 10
12 – 3	15 – 5	14 – 5	12 – 5	15 – 6	19 – 9	17 – 8
11 – 3	12 – 4	16 – 6	19 – 10	18 – 8	14 – 6	16 – 8
8 – 1	15 – 7	18 – 9	17 – 7	11 – 2	10 – 3	13 – 5

3 Use this code to find the message.

A	E	H	I	K	L	M	S	T
8	6	2	3	9	4	5	7	10

10 – 7 = ☐

Ask:
7 plus what makes 10?

7 + ☐ = 10

Message:

10–7	12–8	9–6	11–2	10–4	8–3	9–1	10–0	7–5	9–2

FUN SPOT

Make a message of your own using the code.

 • *AUSTRALIAN SIGNPOST MATHS NSW 2* • ISBN 9780655709039

8B Differences

The difference between 3 and 7 is the same as between 5 and 9.

$7 - 3 = 9 - 5$

CONCEPT

A Is the difference between 11 and 6 the same as between 8 and 3? yes

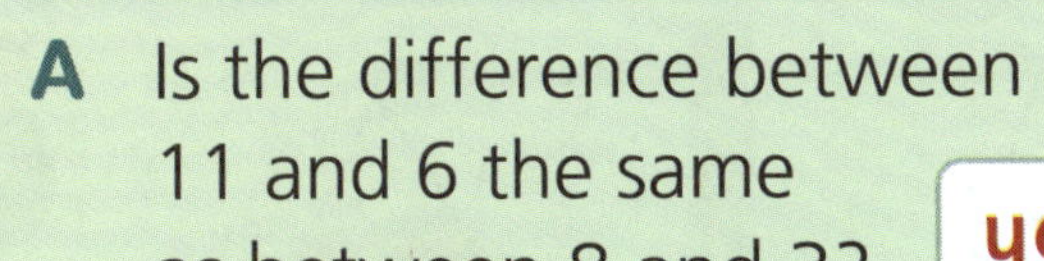

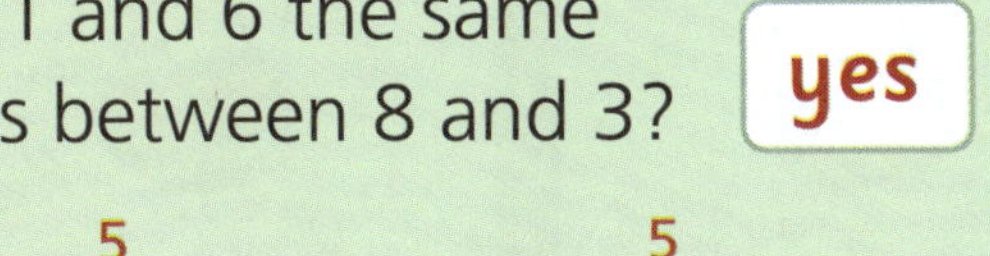

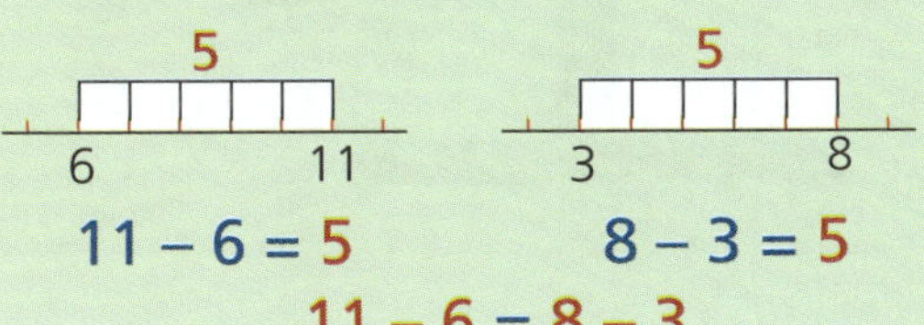

B Finding the difference

The difference between 9 and 2 = 7

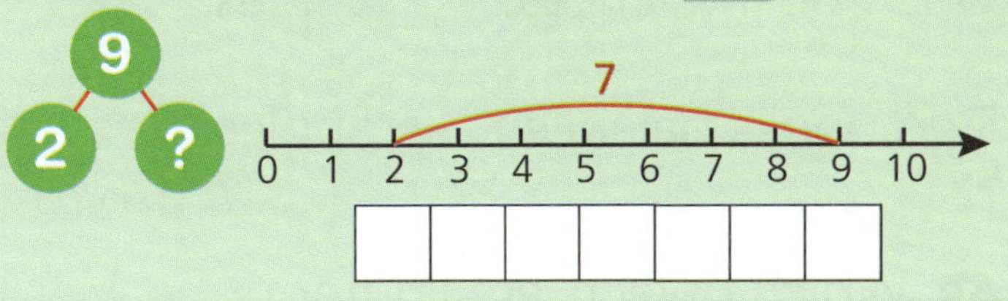

1

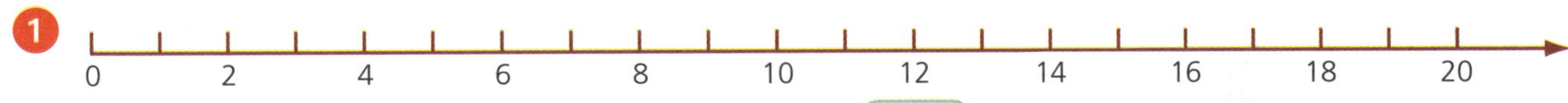

a The difference between 10 and 6 = ☐

b The difference between 15 and 8 = ☐

c The difference between 12 and 6 = ☐

2 **a** Is the difference between 7 and 4 the same as between 17 and 14? ☐

b Is the difference between 8 and 5 the same as between 16 and 13? ☐

c Is the difference between 12 and 8 the same as between 11 and 5? ☐

d Is the difference between 19 and 17 the same as between 8 and 6? ☐

3 Fill in the diagram using addition facts.

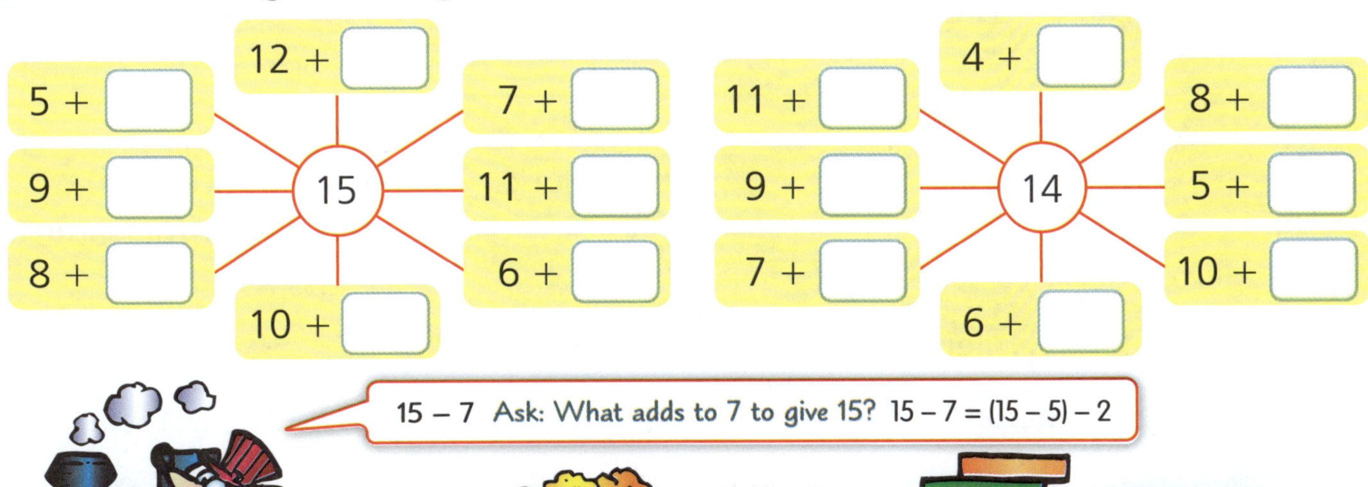

 • *AUSTRALIAN SIGNPOST MATHS NSW 2* • ISBN 9780655709039

8C Balance scales

Use hefting to estimate.

CONCEPT

Using scales is like hefting.

These scales are **balanced**.

One book balances ☐ apples.

Two books balance ☐ apples.

1. Which objects are lighter or heavier than a maths textbook?

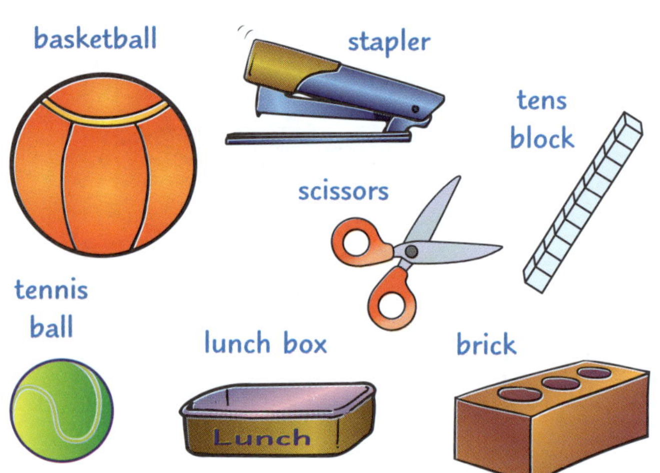

Lighter	Heavier

ACTIVITY

- Estimate, then balance and record.

Object	Unit of measure
cup	☐ tens blocks
scissors	☐ tens blocks
ruler	☐ tens blocks

- Which of the three objects was heaviest?

- Find an object that has nearly the same mass as a cup.

- Would you need more ones blocks or tens blocks to balance a cup? ☐ Discuss why.

Comparing masses

Why would I need more ones blocks than marbles?

ACTIVITY

1 a Handle two objects like these and **estimate** which one is heavier.

I estimate the ______ to be heavier than the ______.

pencil case   lunch box

b Use blocks and a balance scale to find the mass of each object. (You could also use marbles as a unit.)

The ______ has a mass equal to ______ marbles.

The ______ has a mass equal to ______ marbles.

2 Choose three pairs of objects in the classroom.

- Estimate which object is heavier in each pair by hefting.
- Check by using a balance scale and marbles.
- Complete this chart.

Estimate by hefting	Number of units	Was your estimate correct?
______ is heavier than ______	______ ______	☐ Yes ☐ No
______ is heavier than ______	______ ______	☐ Yes ☐ No
______ is heavier than ______	______ ______	☐ Yes ☐ No

INVESTIGATION

Estimating mass

Choose two objects. Estimate what their mass would be using marbles. Use a balance scale to measure. How close were your estimates?

Object	Estimate	Real mass	Difference
	marbles	marbles	marbles
	marbles	marbles	marbles

9A Linking addition and subtraction

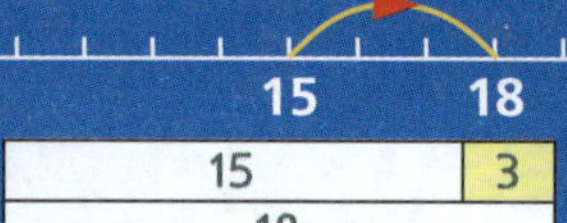

CONCEPT

If 8 + 12 = 20 then 12 + 8 = 20

20 − 8 = 12

20 − 12 = 8

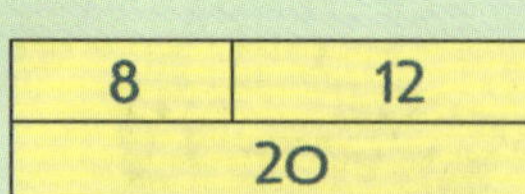

13 + 6 = 19

6 + 13 = 19

19 − 6 = 13

19 − 13 = 6

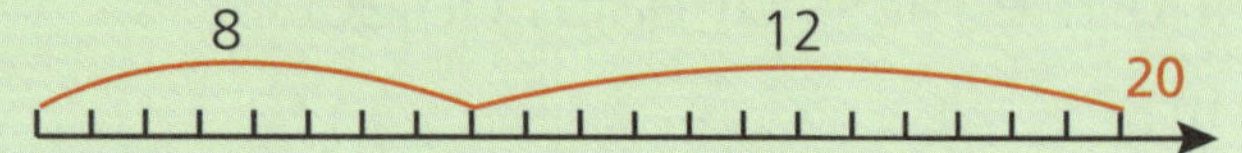

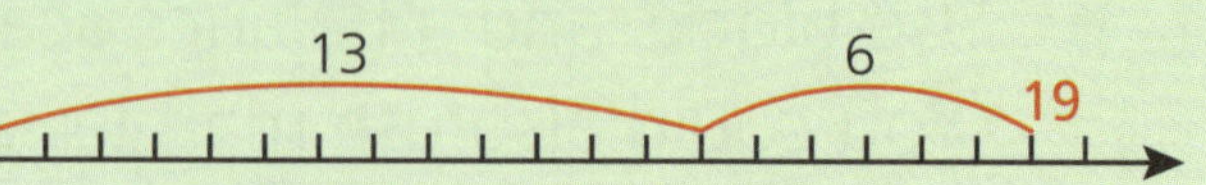

1 **a** If 7 + 15 = 22 then 15 + 7 = ☐

22 − 7 = ☐

22 − 15 = ☐

7 + 15 = ☐

b 9 + 8 = ☐

8 + 9 = ☐

17 − 9 = ☐

17 − 8 = ☐

2 Use the pictures to help fill in the missing numerals.

a

☐ + ☐ = ☐

☐ + ☐ = ☐

☐ − ☐ = ☐

☐ − ☐ = ☐

b

☐ + ☐ = ☐

☐ + ☐ = ☐

☐ − ☐ = ☐

☐ − ☐ = ☐

3 Write two addition and two subtraction facts for this picture.

☐ + ☐ = ☐ ☐ − ☐ = ☐

☐ + ☐ = ☐ ☐ − ☐ = ☐

4 Write two addition and two subtraction facts for this picture.

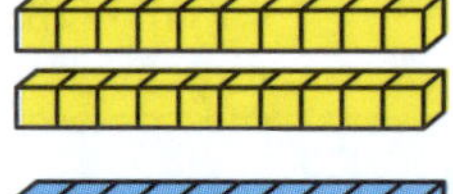

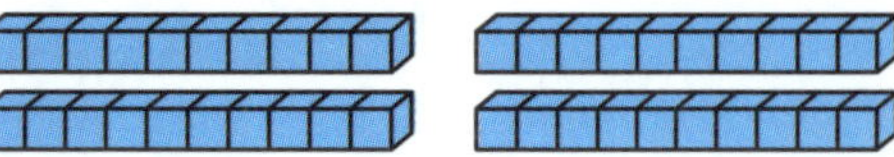

☐ + ☐ = ☐ ☐ − ☐ = ☐

☐ + ☐ = ☐ ☐ − ☐ = ☐

 • *AUSTRALIAN SIGNPOST MATHS NSW 2* • ISBN 9780655709039

9B Linking addition and subtraction

12	8
20	

CONCEPT

6 + 4 = 10

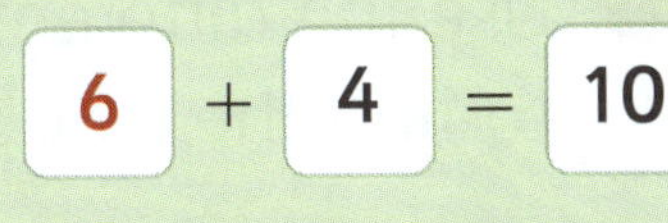

so 10 − 6 = 4 and 10 − 4 = 6

12 + 8 = 20 → 20 − 8 = 12; 20 − 12 = 8

1 Write two linking subtraction number sentences for each addition.

a 8 + 7 = 15 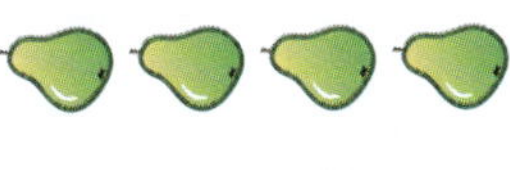  ___ − ___ = ___ ; ___ − ___ = ___

b 6 + 8 = 14 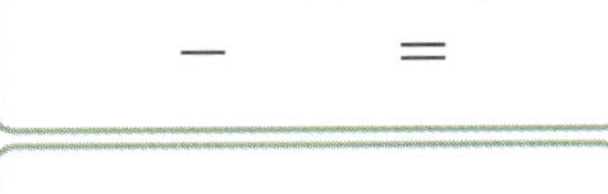 ___ − ___ = ___ ; ___ − ___ = ___

c 7 + 5 = 12   ___ − ___ = ___ ; ___ − ___ = ___

d 14 + 6 = 20   ___ − ___ = ___ ; ___ − ___ = ___

e 22 + 9 = 31  ___ − ___ = ___ ; ___ − ___ = ___

f 17 + 9 = 26   ___ − ___ = ___ ; ___ − ___ = ___

FUN SPOT

Make linked number sentences of your own.

☐ + ☐ = ☐ ☐ − ☐ = ☐ ☐ − ☐ = ☐

 • *AUSTRALIAN SIGNPOST MATHS NSW 2* • ISBN 9780655709039

9C Informal units of length

CONCEPT

How many times will the width of my finger fit along my pencil?

[ ] finger widths

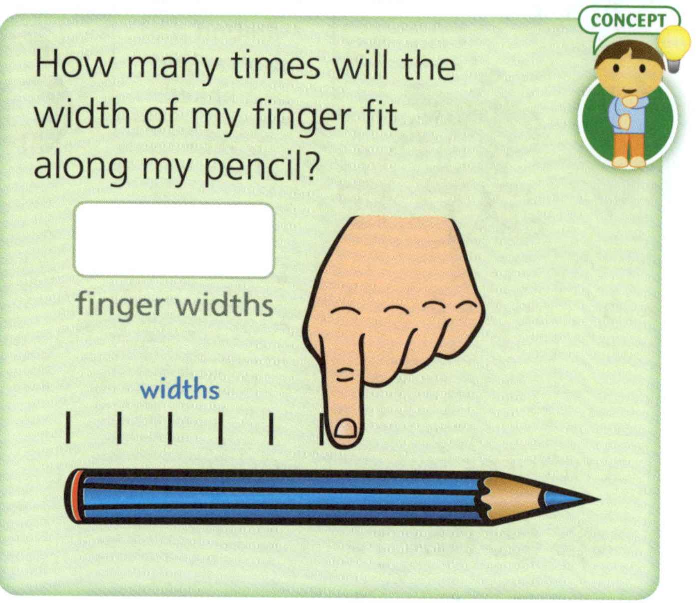

1 Use the different measuring units to find the length of your pencil.

Unit used	Length of your pencil
place-value ones blocks	[ ] blocks
paperclips	[ ] paperclips
finger widths	[ ] finger widths

2 Use hand spans to measure the length of:

	Guess	Check
a this book		
b a bag		
c a window		
d your arm.		

Order these lengths from shortest to longest.

ACTIVITY

Estimate then measure how many steps from where you are to:

	Guess	Check
a the school canteen		
b the lunch seats		
c the library.		

Each step should be the same length.

Order these distances from shortest to longest.

9D Informal units of length

1. Use the different measuring units to find the length of your desk.

Unit used	Length of your desk
this book	☐ books
a pencil	☐ pencils
finger lengths	☐ fingers

2. Use your hand span to measure the length of each object.

Your desk	☐ hand spans
Teacher's desk	☐ hand spans

Which is longer? ☐

Cupboard	☐ hand spans
Door	☐ hand spans

Which is shorter? ☐

3. Estimate, then use craft sticks to measure these lengths.

a 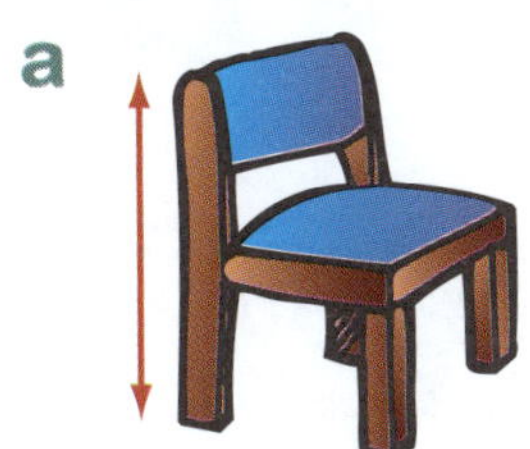

Estimate ☐ sticks
Measure ☐ sticks

b 

Estimate ☐ sticks
Measure ☐ sticks

c 

Estimate ☐ sticks
Measure ☐ sticks

ACTIVITY

Use string to compare the length of objects in the room.

☐ is longer than ☐.

☐ is longer than ☐.

10A How many more?

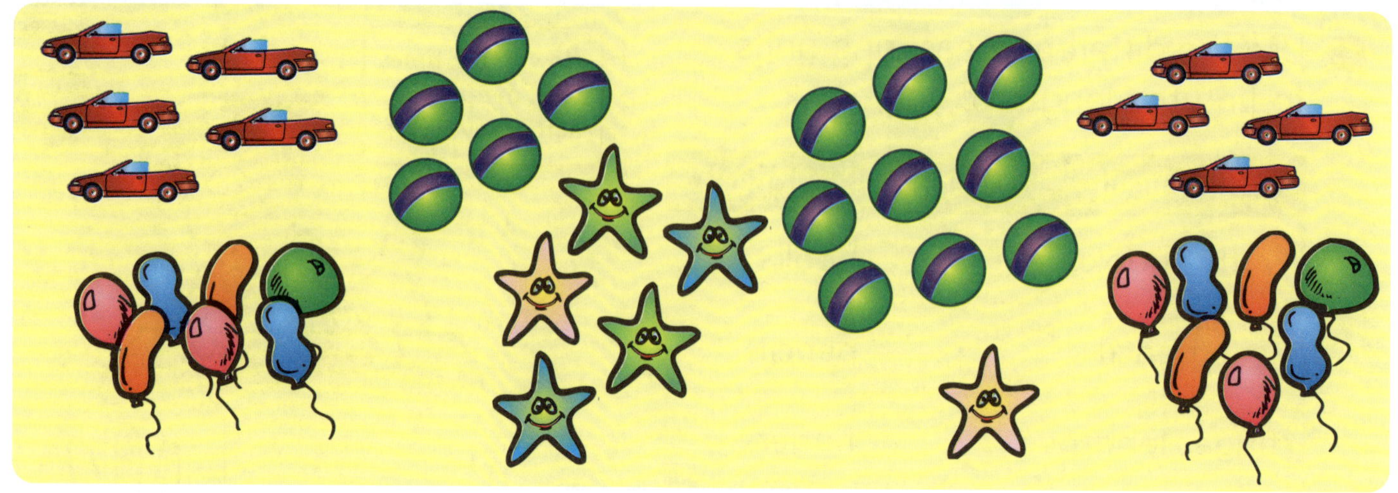

1 In this picture, how many more would I need to make 20:

a cars? 5 + 4 + ___ = 20

b balloons? ___ + ___ + ___ = 20

c balls? ___ + ___ + ___ = 20

d starfish? ___ + ___ + ___ = 20

2
a 6 + ___ = 12

b 3 + 8 + ___ = 16

c 11 + ___ = 15

d 4 + ___ + 5 = 13

e ___ + 9 = 13

f 10 + ___ + 5 = 19

g 5 + ___ = 12

h 8 + 9 + ___ = 20

i ___ + 8 = 16

j 7 + 3 + ___ = 20

Describe how you found the missing number.

INVESTIGATION

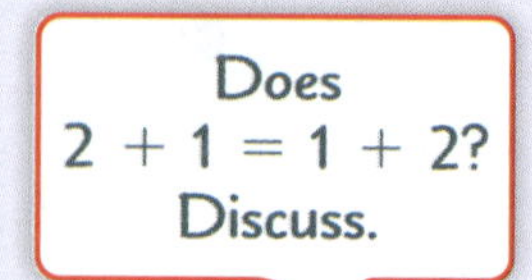

- List all possible answers.

a
6 + ___ + ___ = 10
6 + ___ + ___ = 10
6 + ___ + ___ = 10
6 + ___ + ___ = 10
6 + ___ + ___ = 10

b
17 + ___ + ___ = 20
17 + ___ + ___ = 20
17 + ___ + ___ = 20
17 + ___ + ___ = 20

Does 2 + 1 = 1 + 2? Discuss.

- Explore possible answers for 10 + ___ + ___ = 20.

Volume and capacity

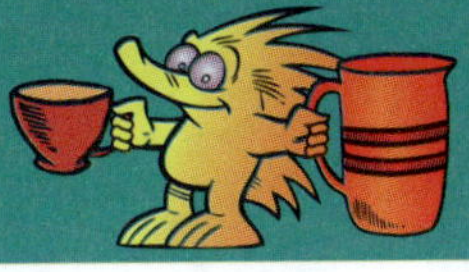

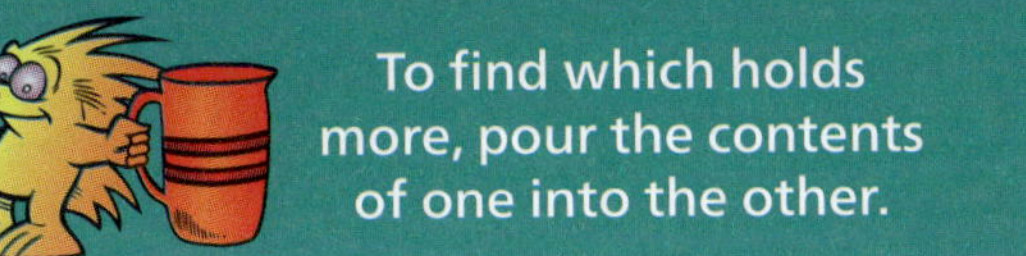

To find which holds more, pour the contents of one into the other.

1 Dipping into water:

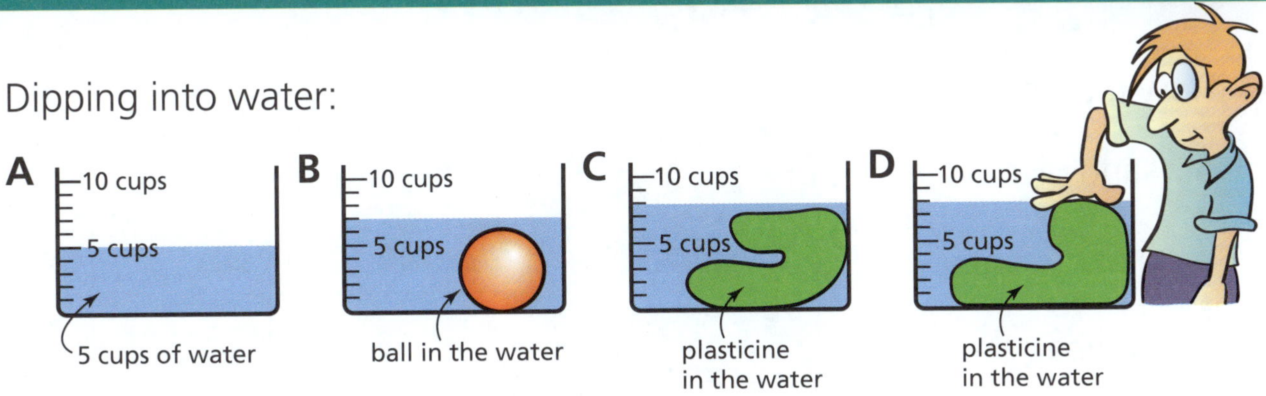

A 10 cups, 5 cups — 5 cups of water
B 10 cups, 5 cups — ball in the water
C 10 cups, 5 cups — plasticine in the water
D 10 cups, 5 cups — plasticine in the water

a What is the volume of the ball? ☐ cups

b What is the volume of two balls? ☐ cups

c What is the volume of the plasticine in C? ☐ cups

d What is the volume of the plasticine in D? ☐ cups

e Could the plasticine in C and D be the same? ☐

2 Look at the diagram.

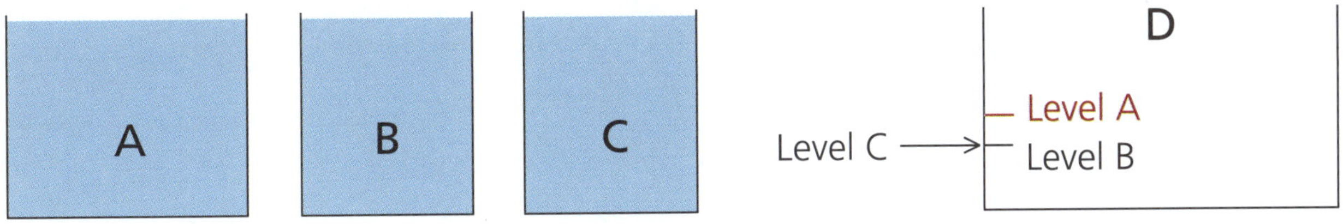

A B C D

Level A
Level C → Level B

a If the water in A is poured into B, would it overflow? ☐

When the water from A is poured into D it reaches Level A.

b Does container B have the same capacity as container A? ☐

c Which container has the same capacity as container C? ☐

d Which container has the greatest capacity? ☐

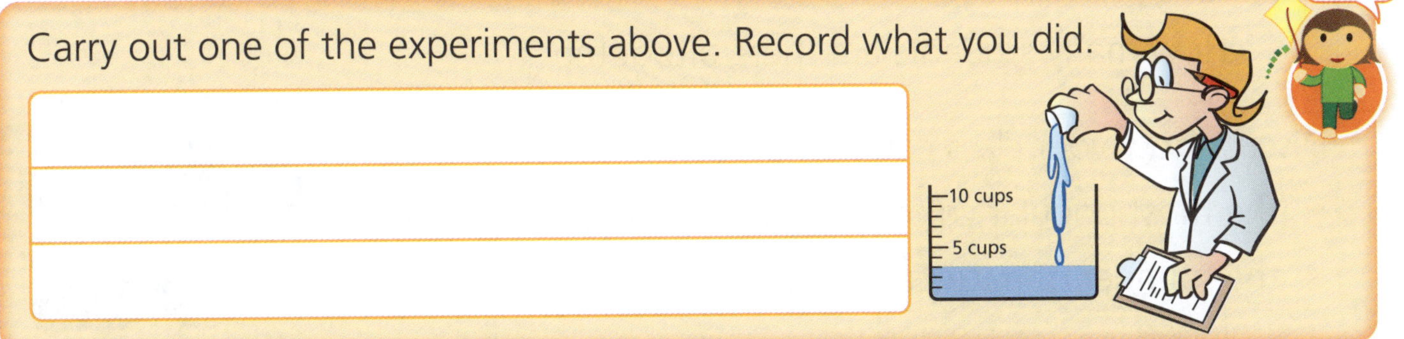

Carry out one of the experiments above. Record what you did.

Statistics and probability

10C Using graphs

1 How many:

a sandwiches?

b pies?

c apples?

d altogether?

e more pies than apples?

f Complete the table below to show the food in the classroom.

Food in the classroom	
Sandwiches	
Pies	
Apples	

g What other foods could be in a classroom?

2 Each student put a candle above their month of birth to make this graph.

Jan	Feb	Mar	Apr	May	Jun	Jul	Aug	Sep	Oct	Nov	Dec

How many students were born in:

a January? b November? c February?

d How many students are there altogether?

FUN SPOT

Discuss how your class could find out:

- the most popular colours of the cars passing the school.
- the number of each colour in a box of Smarties (or M&Ms).

1 Link each event with the better label.

a

more likely

less likely

b

more likely

less likely

c

more likely

less likely

2 Complete this sentence.
It is more likely that I will

3 Sam will take out one counter. Match the boxes.

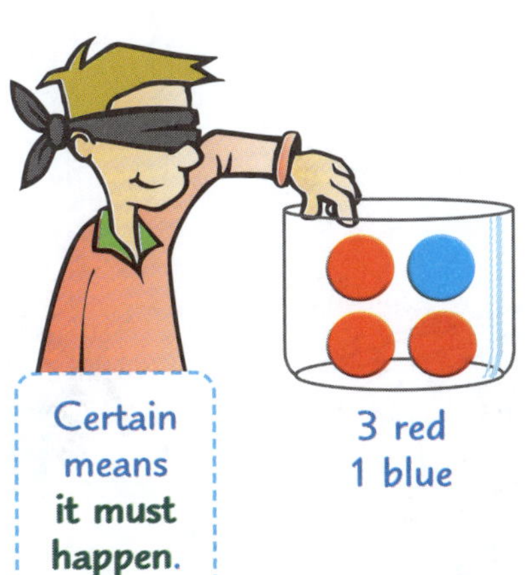

Certain means it must happen.

 red	 yellow	blue	 red or blue

impossible | likely | certain | unlikely

11A Adding 10s

CONCEPT

I added each row to my total.

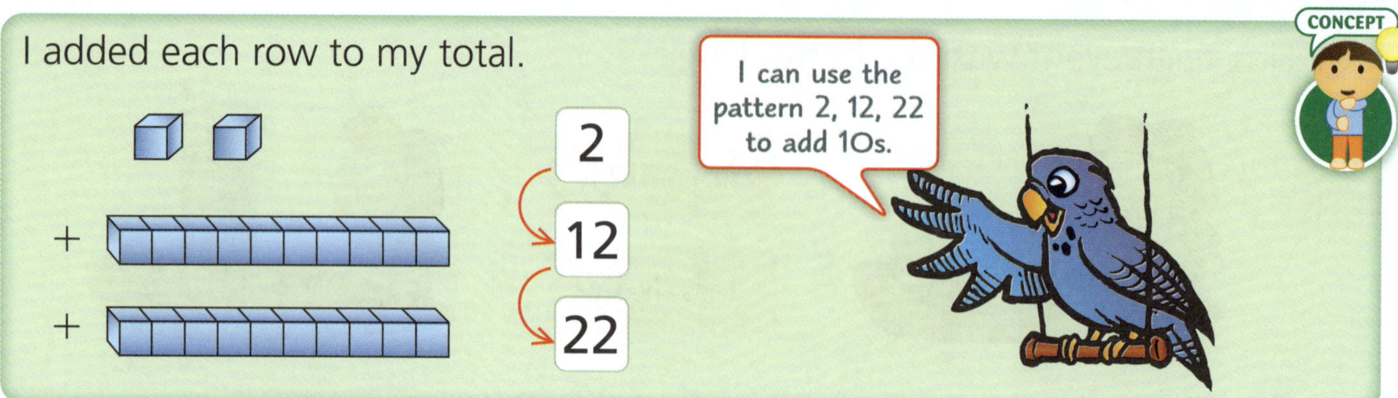

1. Write the totals, adding another 10 for each row.

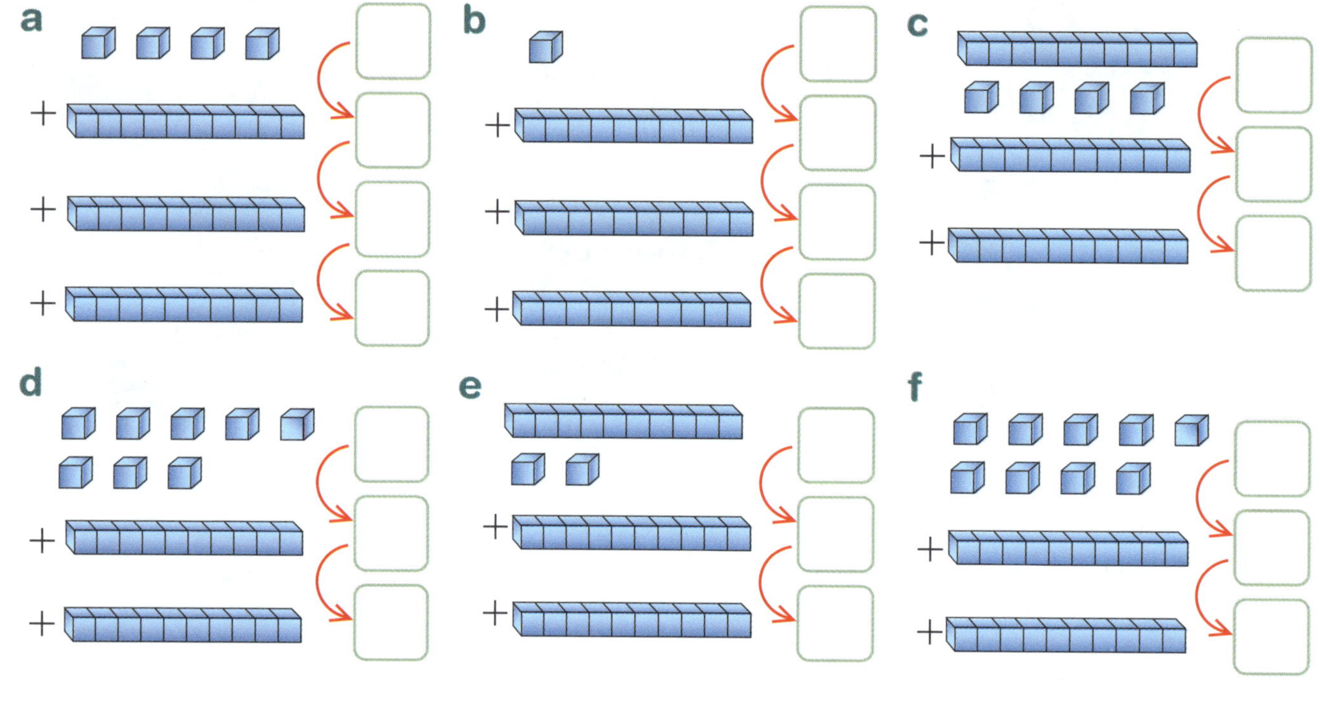

2. Add 20 to each number.

a 3 ☐ b 36 ☐ c 17 ☐ d 52 ☐ e 94 ☐

3. Add 30 to each number.

a 1 ☐ b 35 ☐ c 68 ☐ d 19 ☐ e 33 ☐

ACTIVITY

Line up place-value blocks to create your own addition of tens pattern.

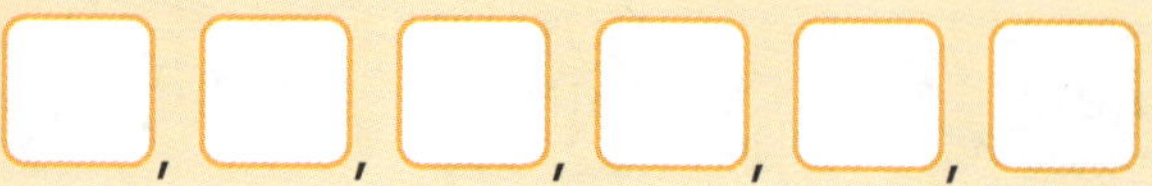

 • *AUSTRALIAN SIGNPOST MATHS NSW 2* • ISBN 9780655709039

11B Adding and subtracting 10s

+ 10
45 + 10 = 55

– 10
45 – 10 = 35

We can use the columns to add 10s or take away 10s.

31, 21 — 31 – 10 = 21

9, 19, 29, 39 — 9 + 30 = 39

43, 33, 23, 13 — 43 – 30 = 13

1	2	3	4	5	6	7	8	9	10
11	12	13	14	15	16	17	18	19	20
21	22	23	24	25	26	27	28	29	30
31	32	33	34	35	36	37	38	39	40
41	42	43	44	45	46	47	48	49	50

1. Help Banjo find the answers by adding or subtracting 10.

2. Continue these patterns.

   a 15, 25, 35, ☐, ☐, ☐

   b 59, 49, 39, ☐, ☐, ☐

   c 37, 47, 57, ☐, ☐, ☐

   d 81, 71, 61, ☐, ☐, ☐

3. Use 10s patterns to complete these.

   a 12 + 20 → 12, 22, 32 so 12 + 20 = ☐

   b 27 + 30 → 27, 37, 47, 57 so 27 + 30 = ☐

   c 51 – 20 → 51, 41, 31 so 51 – 20 = ☐

   d 85 – 30 → 85, 75, 65, 55 so 85 – 30 = ☐

   e 43 + 40 = ☐ **f** 74 – 40 = ☐ **g** 38 + 40 = ☐

   h True or false? 34 + 30 = 94 – 30 ☐

11C Ordering capacities

1 Choose sets of three containers and measure the capacity of each. Record your results in the table below.

a

Container	Estimate	Count
	cups	cups
	cups	cups
	cups	cups

b

Container	Estimate	Count
	cups	cups
	cups	cups
	cups	cups

2 Order each set of containers above from "holds least" to "holds most".

a 1 ______ 2 ______ 3 ______

b 1 ______ 2 ______ 3 ______

3 Tina recorded these results. Answer the following questions.

Container	Number	Comment
large pot	15 cups	
bucket	11 cups	holds least
box	26 cups	holds most

a Which container held 15 cups? ______

b Which container held 26 cups? ______

c Which container held less than the large pot? ______

d Which container held more than the large pot? ______

e Order the containers from smallest to largest.

1 ______ 2 ______ 3 ______

Displaying information

We can show data using objects, pictures, words or symbols.

Table

Pet	Number
Cat	2
Dog	5
Rat	1

Graph: Class pets

Cat	Cat			
Dog	Dog	Dog	Dog	Dog
Rat				

List

Dog, Cat, Dog, Dog, Rat, Cat, Dog, Dog

We might ask why there were only 8 pets.

Use the data above to answer these questions.

a How many cats? ☐

b How many more dogs than cats? ☐

1 Students were asked which animal they liked best out of **elephant**, **lion** and **monkey**. Use the list to finish the graph and table.

List

Monkey, Lion, Lion, Monkey, Elephant, Lion, Monkey, Elephant, Elephant, Elephant, Lion, Elephant, Monkey, Elephant, Monkey

Animal	Number
Elephant	
Lion	
Monkey	

Discuss the results. Could some students have a different favourite?

  • ISBN 9780655709039

12A Inverse operations

CONCEPT

Adding 5 is the opposite of taking away 5.

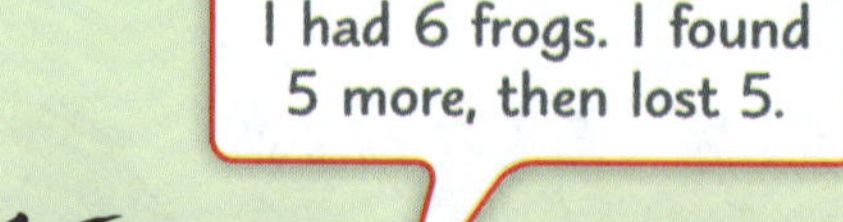

Use counters to act out each problem.

1 Rand had 10 balls. He found 6 more, then gave 6 away.
How many does he have now?

$10 + 6 - 6 = \square$

2 Sarah had 12 stickers. She gave away 5, then was given 5 more.
How many does she have now?

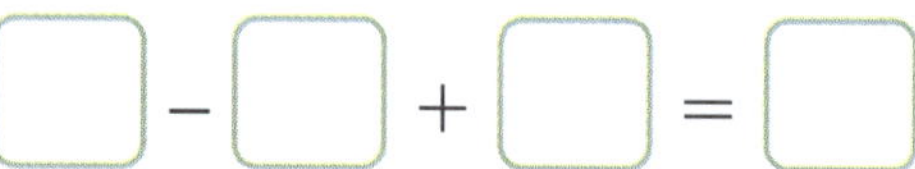

3 Perrin could see 12 dogs. Another 4 dogs came, then 4 left.
How many can he see now?

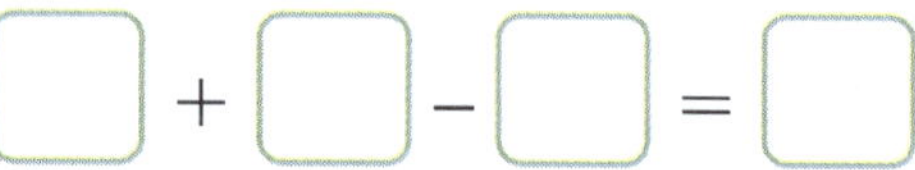

4
- **a** $12 + 3 - 3 = \square$
- **b** $14 + 5 - 5 = \square$
- **c** $13 + 5 - 5 = \square$
- **d** $17 + 7 - 7 = \square$
- **e** $9 + 11 - 11 = \square$
- **f** $7 + 12 - 12 = \square$
- **g** $23 - 8 + 8 = \square$
- **h** $15 - 6 + 6 = \square$
- **i** $21 - 4 + 4 = \square$

ACTIVITY

Create your own number stories and sentences where the addition and subtraction undo each other.

 • *AUSTRALIAN SIGNPOST MATHS NSW 2* • ISBN 9780655709039

12B Australian money

1 Write the value of each coin.

2 Match each banknote to the correct label.

$20
twenty dollars

$50
fifty dollars

$100
one hundred dollars

$10
ten dollars

$5
five dollars

3 Write the value of the banknotes in order from smallest to largest.

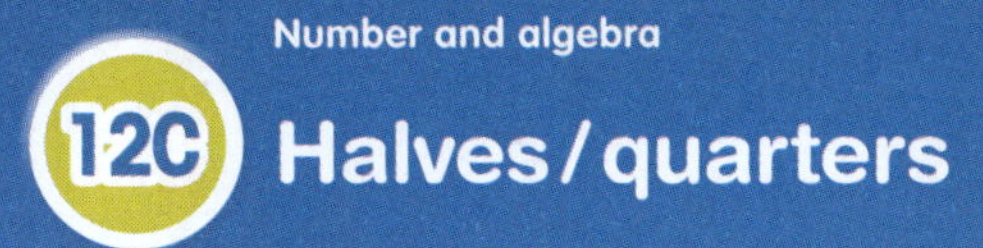

12C Halves / quarters

12			
6		6	
3	3	3	3

whole
halves
quarters

Jake lined up 8 apples.

CONCEPT

8	
4	4

He put a pencil at the halfway point.
He shared half of his apples with Mae. Each half has 4 apples.

1. Line up an even number of items. Use a pencil to mark the halfway point.
   - **a** Write what you did.
   - **b** How many in each half? ☐
   - **c** When you take away the pencil, how many items are there? ☐

Toby shared 8 books equally among 4 friends. He put the books in a line.

CONCEPT

He put a pencil at the halfway point.
He then halved each half, placing pencils at those points too.

8			
2	2	2	2

There were 2 books in each quarter. Each friend was given 2 books.

2. Line up 12 counters. Use 3 pencils to separate them into quarters.
   - **a** Write what you did.
   - **b** How many in each quarter? ☐
   - **c** When you take away the pencils, how many counters are there? ☐

Fractions of a group

- The two halves of a collection will have the same number of items.
- To find one quarter of a collection, we halve and halve again.
- Halving again gives us eighths.

1 Circle half of each group.

a 

b 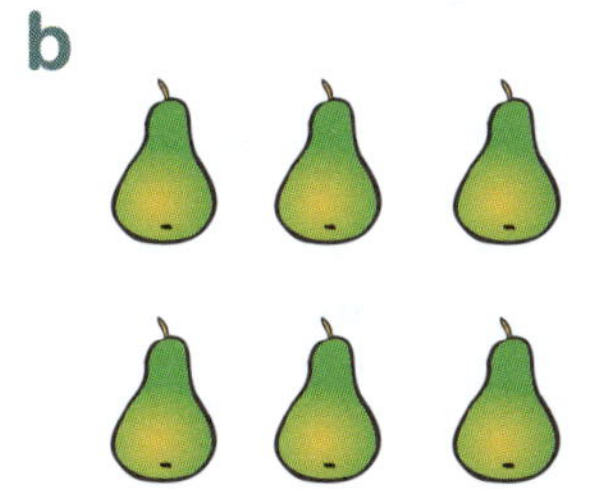

c 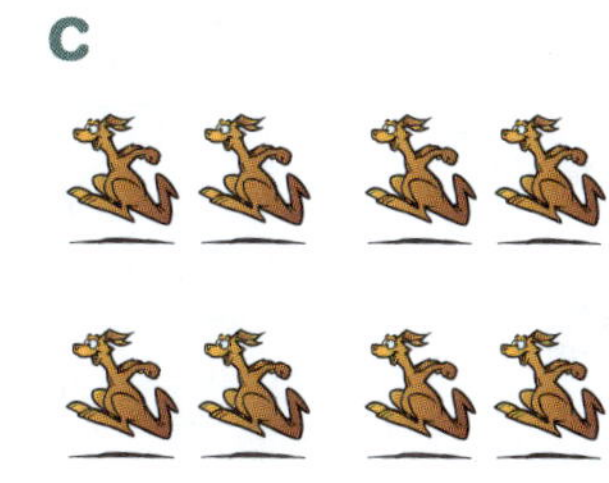

d 

e 

2 Circle one quarter of each group.

a 

b 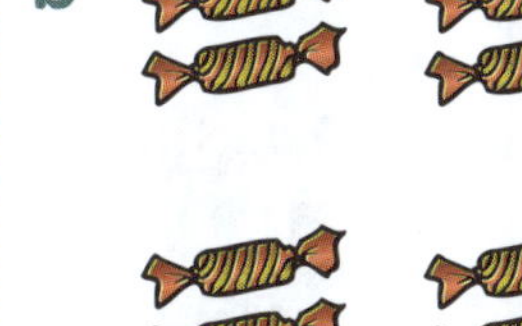

c 

d 

e 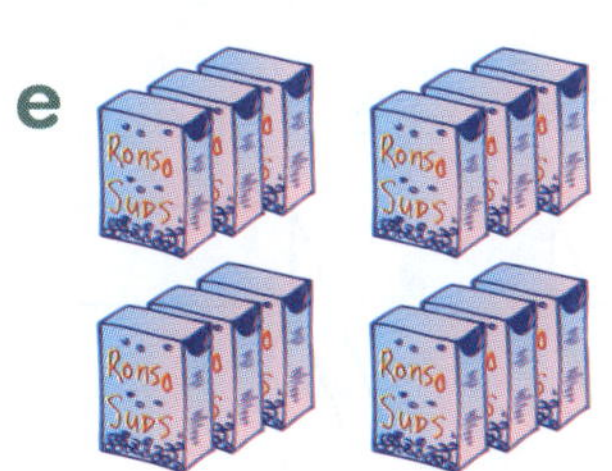

f 

g 

h 

3 Colour one half of each group. Circle one quarter. Underline one eighth.

a 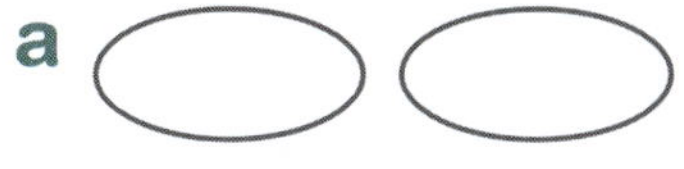 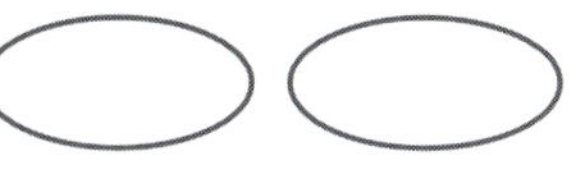 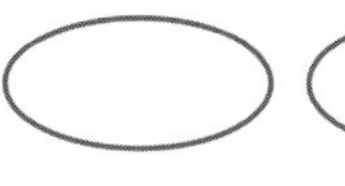 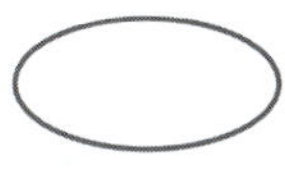  

b 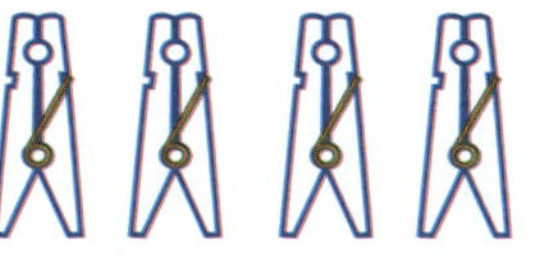 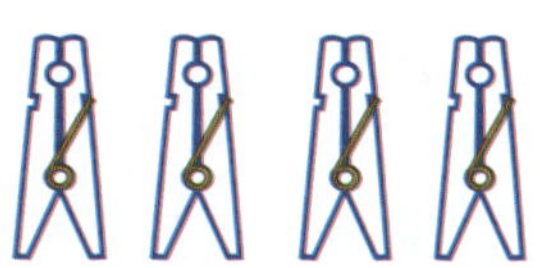 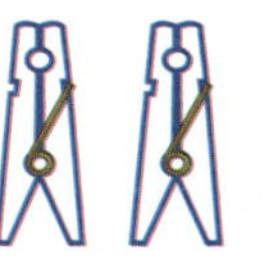

13A Equal groups

four groups of five

5 + 5 + 5 + 5

1 Match each group with the correct number sentence.

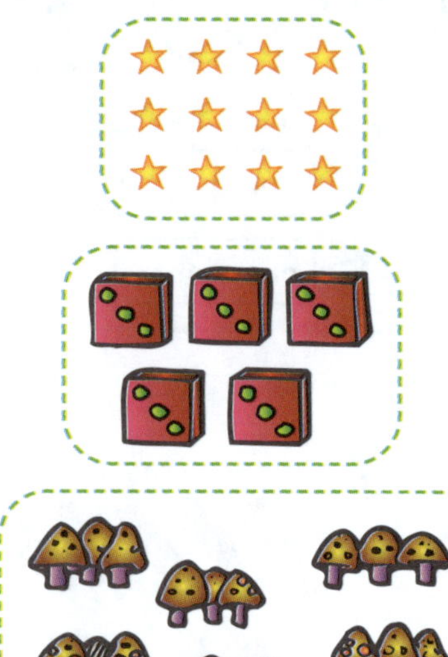

2 rows of 9 = 18
5 groups of 2 = 10
6 groups of 3 = 18
2 groups of 6 = 12
3 rows of 4 = 12
5 groups of 3 = 15

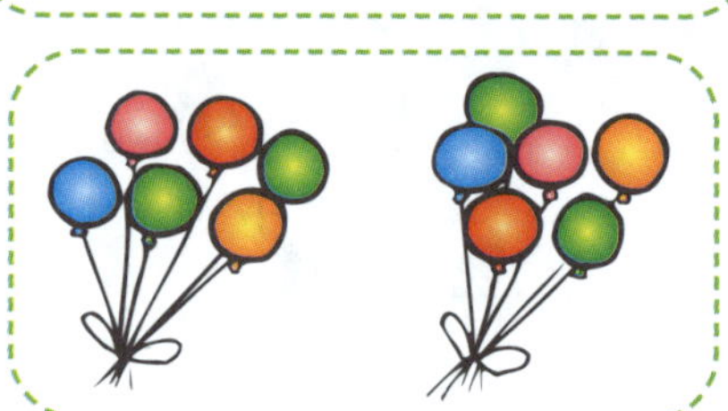

2 Complete these number sentences.

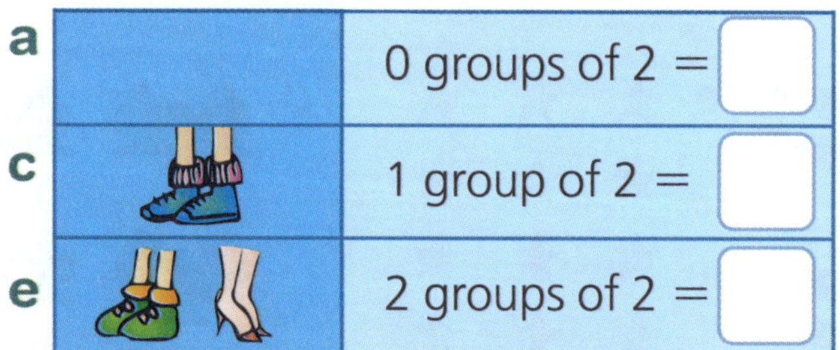

a 0 groups of 2 = ☐

c 1 group of 2 = ☐

e 2 groups of 2 = ☐

b 3 groups of 2 = ☐

d 4 groups of 2 = ☐

f 5 groups of 2 = ☐

3 Draw pictures to show each story. Complete the number sentences.

a 3 vases of 4 flowers each

☐ groups of ☐ = ☐

b 2 boxes with 6 balls in each

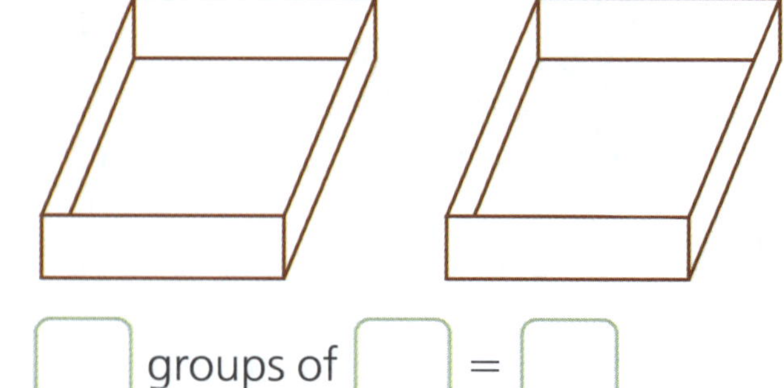

☐ groups of ☐ = ☐

4 Use counters to make arrays to answer these questions.

a 3 groups of 7 = ☐ **b** 4 groups of 6 = ☐ **c** 5 groups of 3 = ☐

 • *AUSTRALIAN SIGNPOST MATHS NSW 2* • ISBN 9780655709039

2 groups of 2 = 4
3 groups of 2 = 6
4 groups of 2 = 8

1 Complete the number sentences.

a

☐ groups of ☐ = ☐

b

☐ groups of ☐ = ☐

c
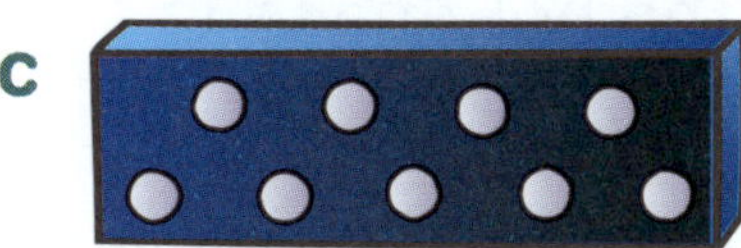

1 group of ☐ = ☐

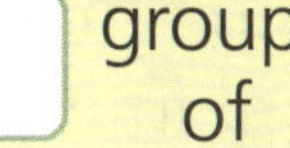
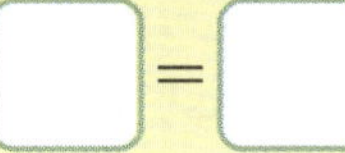

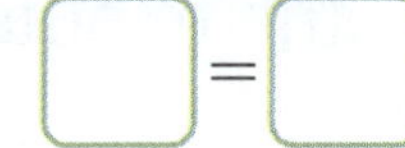

d

☐ groups of ☐ = ☐

e
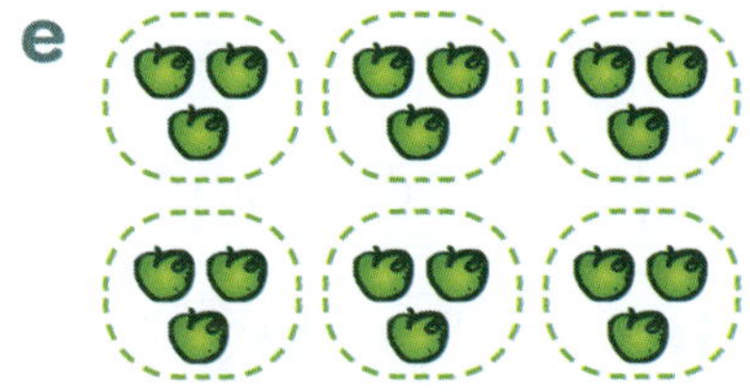

☐ groups of ☐ = ☐

f

☐ groups of ☐ = ☐

g

☐ groups of ☐ = ☐

h

☐ groups of ☐ = ☐

i
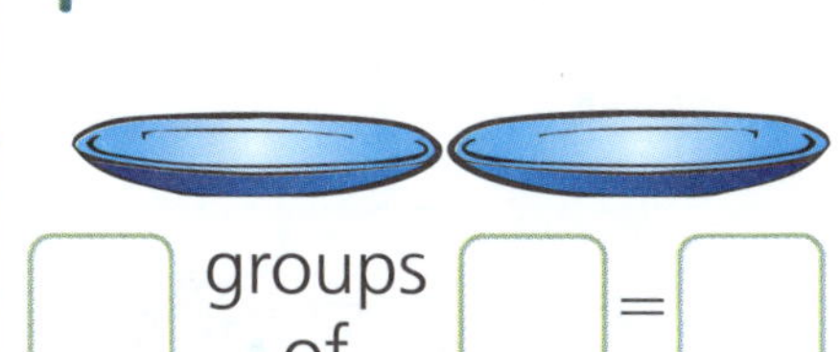

☐ groups of ☐ = ☐

2 Make up your own number sentences and draw pictures to match them.

a ☐ groups of ☐ = ☐

b ☐ groups of ☐ = ☐

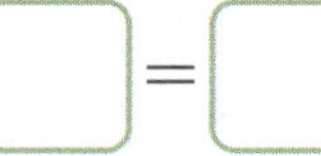

INVESTIGATION

Use the dots to help you complete the number sentences.

2 groups of 6 = ☐

3 groups of 4 = ☐

4 groups of 3 = ☐

6 groups of 2 = ☐

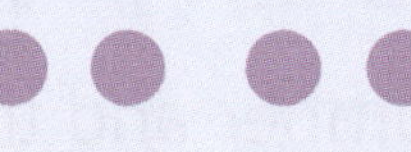

13C Analog time

quarter past 2
2 fifteen
fifteen past 2

quarter to 3
2 forty-five
forty-five past 2

At quarter past, the hound hand is pointing just past the number and the minute hand is pointing to the 3.

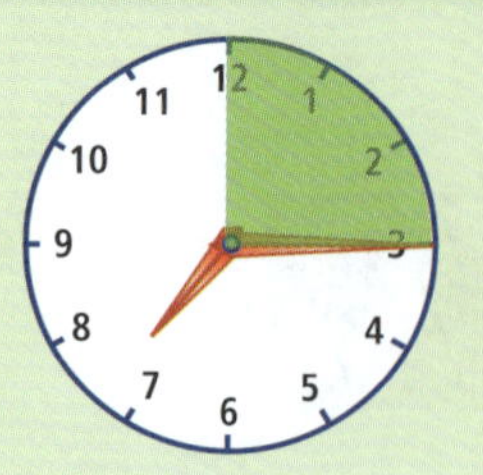

15 past 7
a quarter past 7

At quarter to, the hour hand is almost pointing to the number and the minute hand is pointing to the 9.

45 past 7
a quarter to 8

1 Write the time shown on each face using **quarter to** or **quarter past**.

a 

b 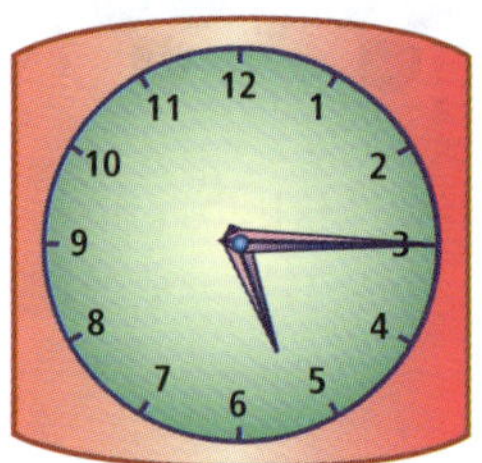

c 

d 

e 

f 

2 On each face show the time given.

a 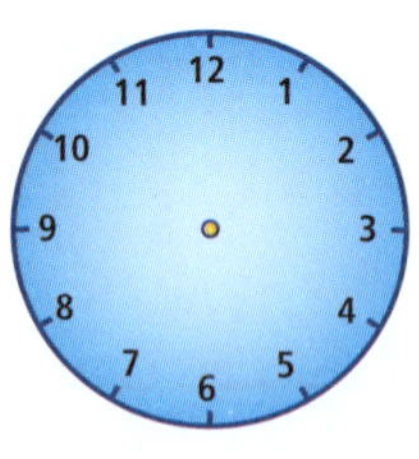
a quarter past 2

b 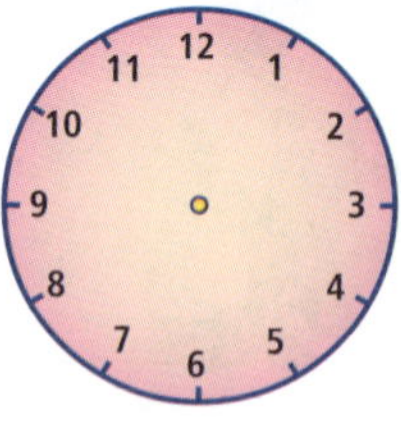
a quarter to 4

c 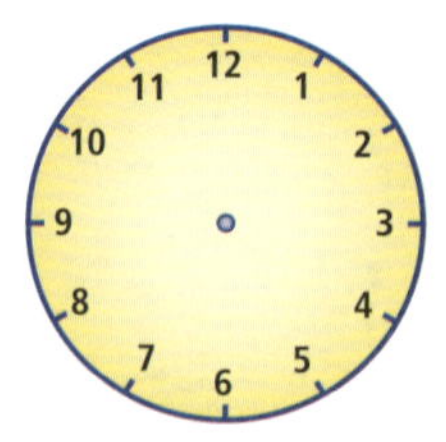
a quarter to 10

Quarter past 10, 15 past 10 or ten fifteen.

3 At a quarter ______ the hour, the small hand is pointing just past the hour number and the big hand is pointing to the ___.

Analog time

quarter past 8	half past 8	quarter to 9
8 fifteen	8 thirty	8 forty-five

CONCEPT

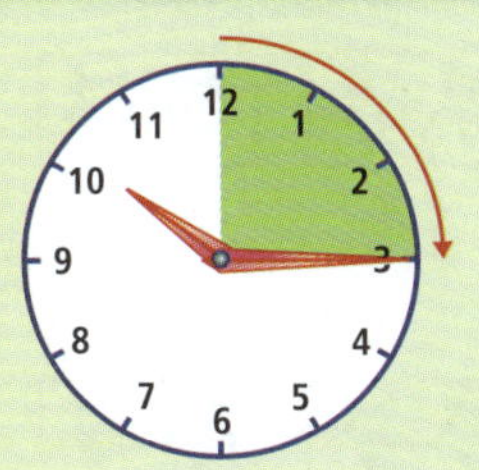

The minute hand has moved a quarter of the way around the clock.

a quarter past 10

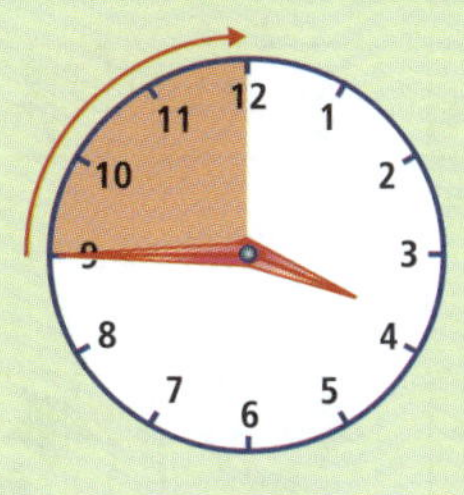

The minute hand has a quarter of the clock to go before it is 4 o'clock.

a quarter to 4

1 Write the time shown.

a 

b 

c 

d 

e 

f 

2 Show the time.

a 

a quarter to 1

b 

a quarter past 7

c 

a quarter to 5

INVESTIGATION

Find how many minutes are in:

a a quarter of an hour

b three quarters of an hour

1 hour = 60 minutes

half an hour = 30 minutes

14A Using skip counting

Can you skip count by 10s?
10, 20, 30, 40, 50, 60 ...

CONCEPT

Skip counting by 2s	2	4	6	8	10	12	14	16 ...
Skip counting by 5s	5	10	15	20	25	30	35	40 ...
Skip counting by 3s	3	6	9	12	15	18	21	24 ...

Learn to skip count by 2s, 3s, 5s and 10s.

1
- **a** 3 groups of 2 ☐
- **b** 6 groups of 2 ☐
- **c** 8 groups of 2 ☐
- **d** 4 groups of 2 ☐
- **e** 7 groups of 2 ☐
- **f** 5 groups of 2 ☐
- **g** 2 groups of 5 ☐
- **h** 4 groups of 5 ☐
- **i** 3 groups of 5 ☐
- **j** 5 groups of 5 ☐
- **k** 8 groups of 5 ☐
- **l** 6 groups of 5 ☐
- **m** 3 groups of 3 ☐
- **n** 5 groups of 3 ☐
- **o** 4 groups of 3 ☐
- **p** 7 groups of 3 ☐
- **q** 6 groups of 3 ☐
- **r** 8 groups of 3 ☐

2 10, 20, 30, 40, 50, 60, 70 ...
- **a** 2 rows of 10 ☐
- **b** 3 rows of 10 ☐
- **c** 5 rows of 10 ☐
- **d** 4 rows of 10 ☐
- **e** 6 columns of 10 ☐
- **f** 7 columns of 10 ☐

4, 8, 12, 16, 20, 24, 28, 32 ...
- **g** 3 columns of 4 ☐
- **h** 5 columns of 4 ☐

You can use skip counting with "rows of" and "columns of" too.

 • *AUSTRALIAN SIGNPOST MATHS NSW 2* • ISBN 9780655709039

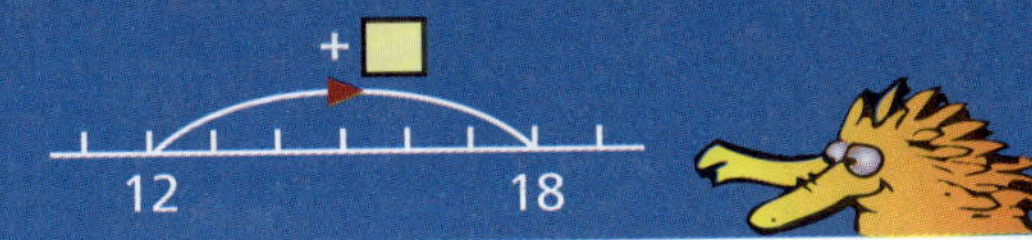

Number line	Number sentence
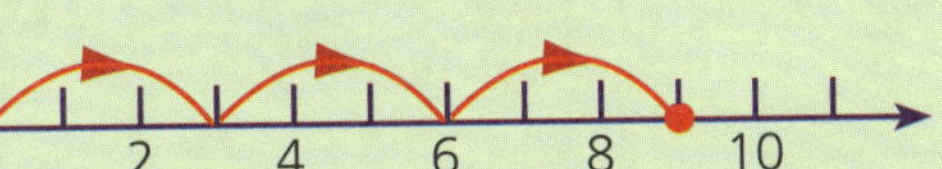	$3 + 3 + 3 = 9$ or 3 groups of $3 = 9$

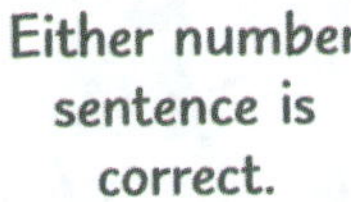

1 Write a number sentence to match each number line.

a 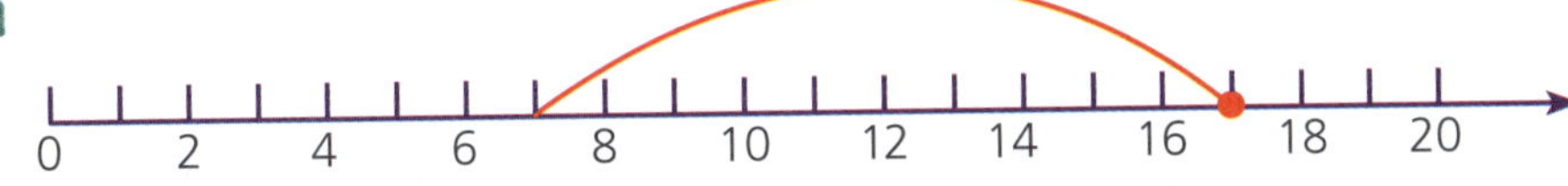

$7 + \square = 17$

b 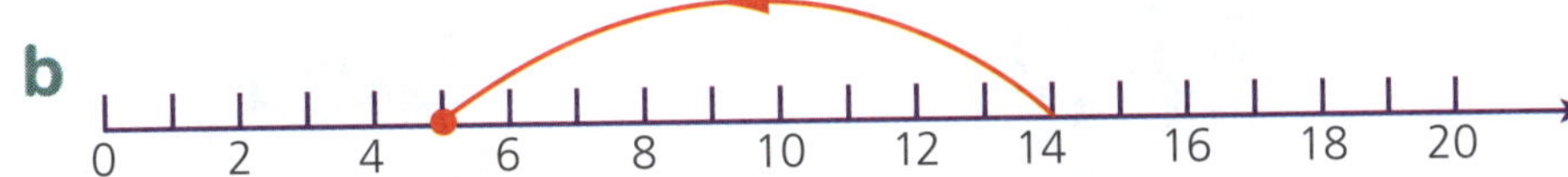

$14 - \square = 5$

c 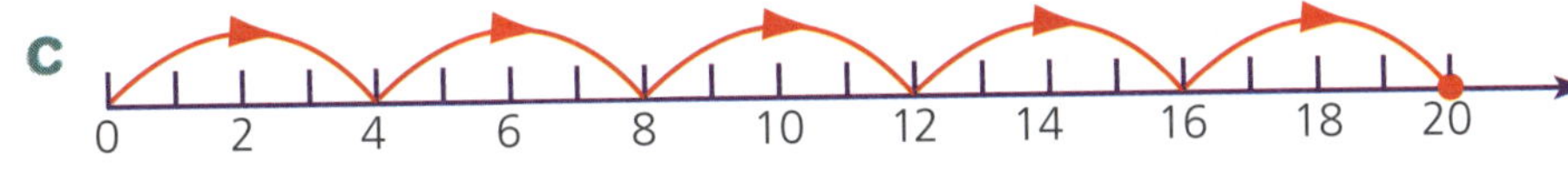

$\square$ groups of $4 = 20$

d 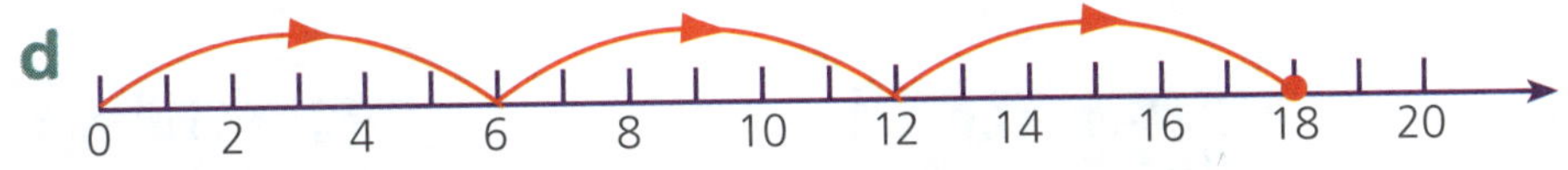

3 groups of $\square = 18$

e 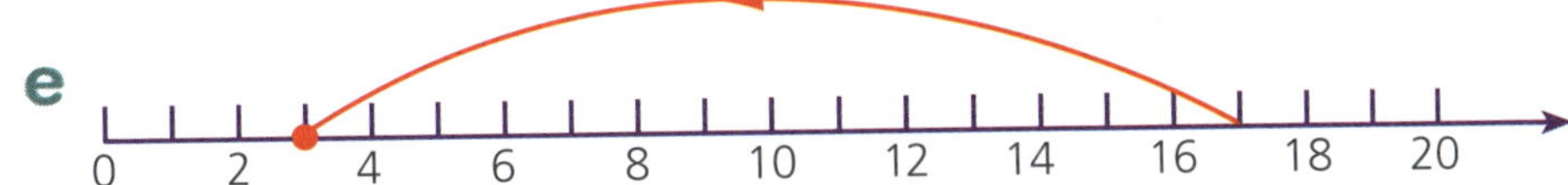

$17 - \square = 3$

f 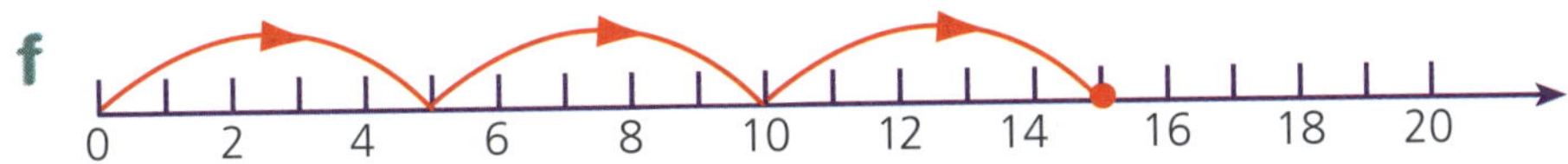

$\square$ groups of $\square = 15$

2 Make up two number sentences and show each on a number line.

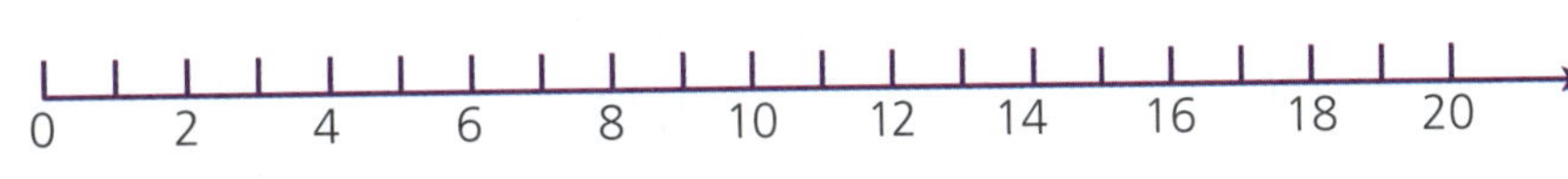

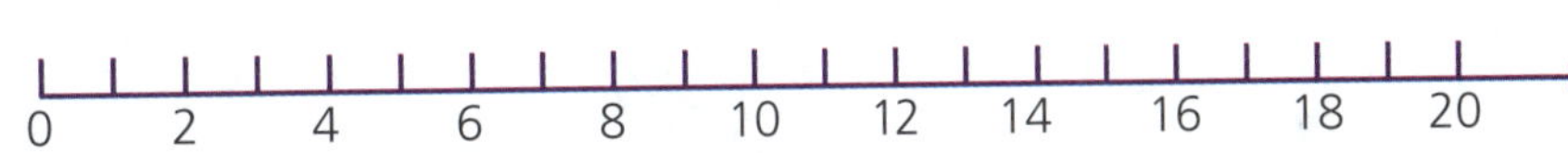

14C Digital time

3 fifteen	3 thirty	3 forty-five
3:15	3:30	3:45
quarter past 3	half past 3	quarter to 4

1 Write two ways we can say the time shown.

a 

or

b 

or

c 

or

d 

or

e 

or

f 

or

2 Write these digital times using numerals and words.

a 7:03 ☐ minutes past ☐

b 9:20 ☐ minutes past ☐

c 4:18 ☐ minutes past ☐

d 3:25 ☐ minutes past ☐

e 6:35 ☐ minutes past ☐

f 2:58 ☐ minutes past ☐

 • *AUSTRALIAN SIGNPOST MATHS NSW 2* • ISBN 9780655709039

14D Analog time

1 Write the time shown on each face.

a

b

c

d

e

f

2 On each face show the time given.

a
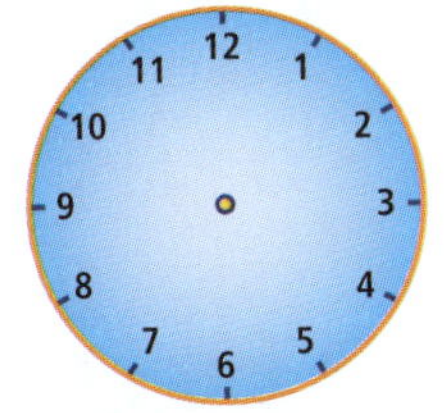
a quarter past 10

b
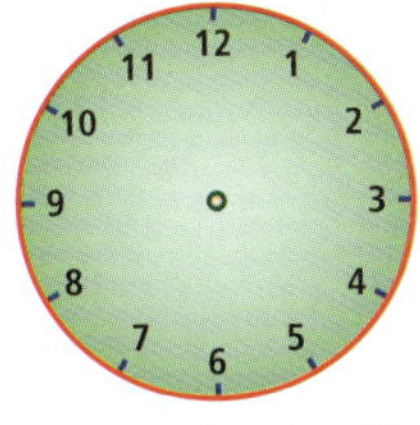
a quarter to 8

c
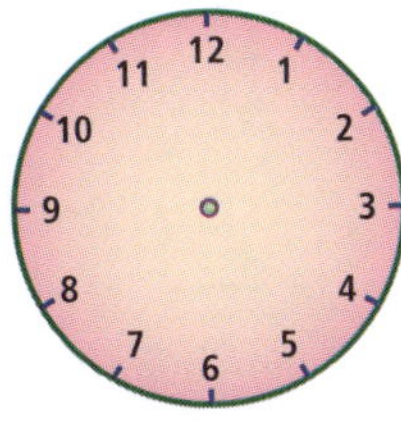
a quarter to 4

d
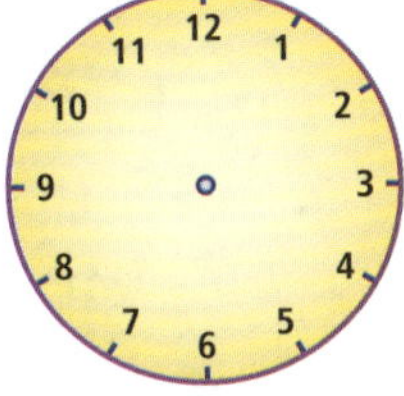
half past 7

e
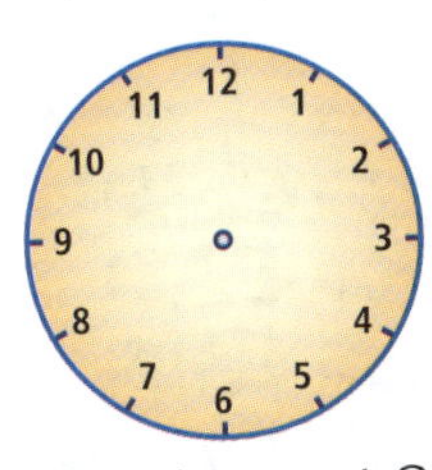
a quarter past 8

f
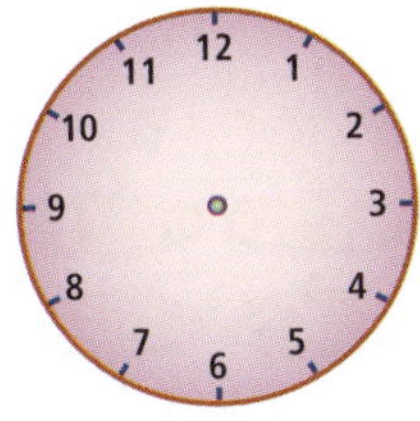
half past 2

g
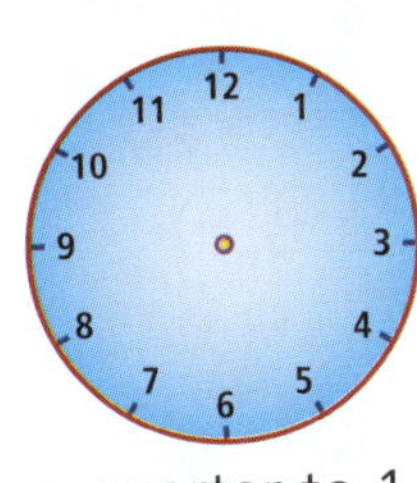
a quarter to 1

h
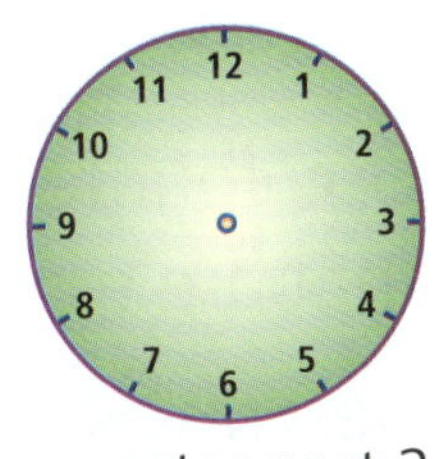
a quarter past 2

15A Using arrays

In arrays, objects are lined up into equal rows and columns.

Felicity's tray held 3 rows of 4 muffins. Cailin's tray held 4 rows of 3 muffins.

3 rows of 4 = 12

4 rows of 3 = 12

Each muffin tray holds 12 muffins.
Felicity turned her tray and found the trays looked the same.

3 rows of 4 = 4 rows of 3 = 12

1 Describe each array and find the total.

a 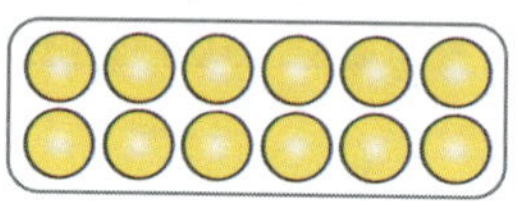

☐ rows of ☐ = ☐

b  

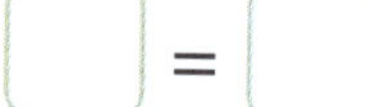

☐ rows of ☐ = ☐

c 2 rows of 6 is the same as ☐ rows of ☐. Total = ☐

d 

☐ rows of ☐ = ☐

e 

☐ rows of ☐ = ☐

f 4 rows of 2 is the same as ☐ rows of ☐. Total = ☐

Use counters to make arrays that are equal.

Describe what you have done.

2 rows of 3 = 6

3 rows of 2 = 6

 • *AUSTRALIAN SIGNPOST MATHS NSW 2* • ISBN 9780655709039

Arrays

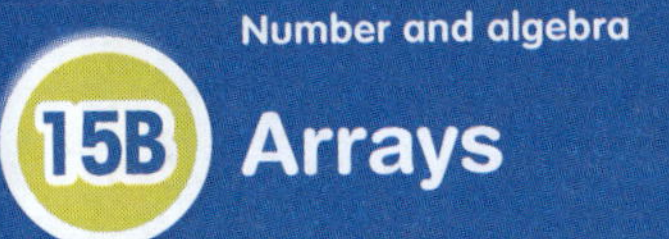

In arrays, objects are lined up into equal rows and equal columns.

CONCEPT

An array is a group of items with **equal rows** and **equal columns**.

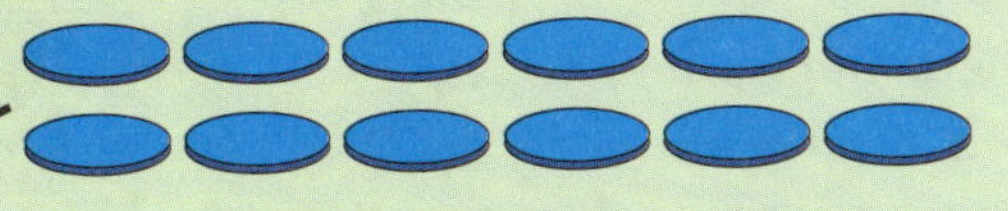

2 rows of 6 = 12
6 columns of 2 = 12

3 rows of 6 = 18
6 columns of 3 = 18

1 Draw and use counters to make these arrays.

a 3 rows of 2 = ☐

b 4 rows of 5 = ☐

c 2 rows of 4 = ☐

2 Draw two of your own arrays and describe them below.

a

rows of =

columns of =

b

3 Colour two or more rectangles. Write two number sentences for each new array.

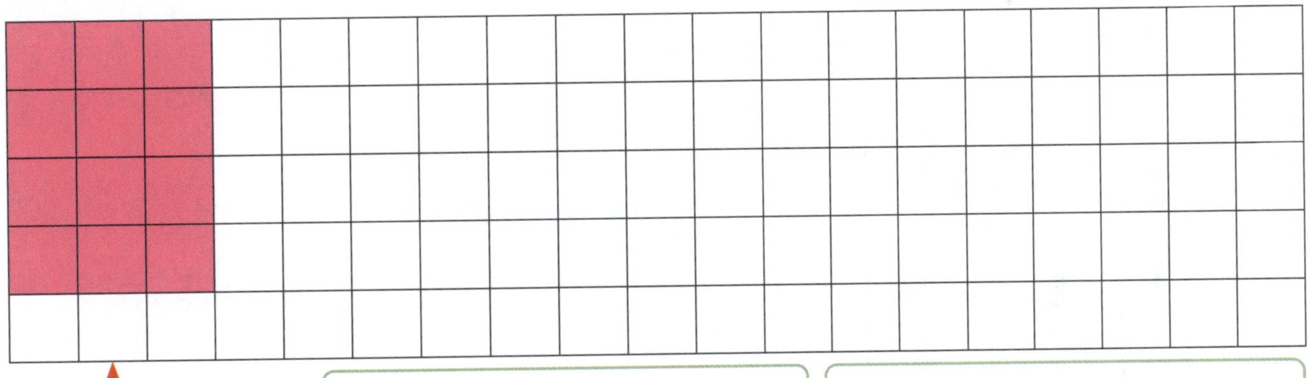

4 rows of 3 = 12
3 columns of 4

 • *AUSTRALIAN SIGNPOST MATHS NSW 2* • ISBN 9780655709039

15C Problem solving

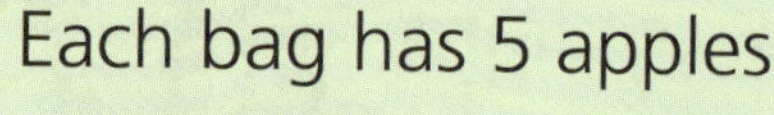

You can skip count by 3.

Each bag has 5 apples.

To find how many apples are in 4 bags, we count the 5 apples four times, 5 + 5 + 5 + 5 or skip count by 5s.

+ + + = 20

1 Use the picture above to find how many apples are in:

a 2 bags ☐

5	5

b 3 bags ☐

5	5	5

c 4 bags ☐

5	5	5	5

2

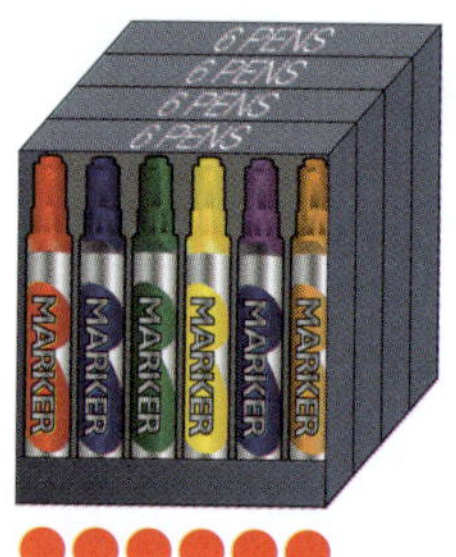

In each packet there are 6 pens.
How many pens are in:

a 2 packets? ☐

b 3 packets? ☐

c 4 packets? ☐

3

In each pod there are 4 peas.
How many peas are in:

a 2 pods? ☐

b 3 pods? ☐

c 4 pods? ☐

d 5 pods? ☐

e 6 pods? ☐

4

In each can there are 3 tennis balls.
How many balls are in:

a 2 cans? ☐

b 3 cans? ☐

c 4 cans? ☐

d 5 cans? ☐

e 6 cans? ☐

f 10 cans? ☐

 ISBN 9780655709039

1 Is it **certain**, **likely**, **unlikely** or **impossible** that a red ball would be chosen?

6 red, 4 blue

3 red, 7 blue

all red

all blue

2 Draw 10 apples. Colour them red or green so that green is more likely to be chosen if a friend chose without looking.

3 Roll a die six times. Write each number rolled.

a Which number was rolled most?

b Which numbers were not rolled?

c What numbers could be rolled on a die?

d Is it possible to throw six ones in a row?

- Choose a number on the die.
- Count how many rolls it takes until your number is rolled. ____ rolls

16A Numbers to 150

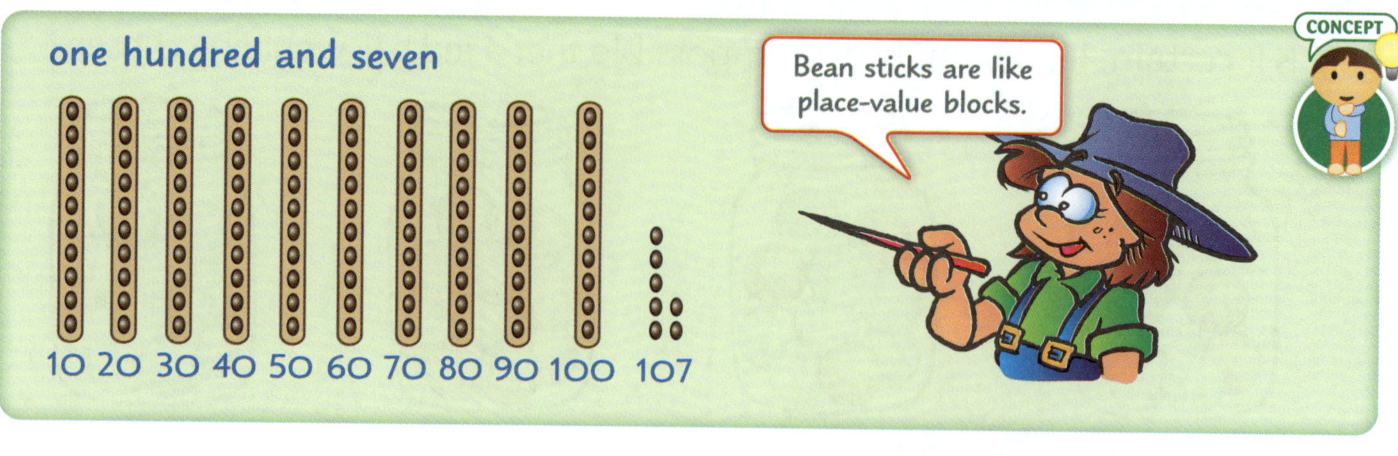

1 Write these numerals.

a [ ] hundreds [ ] tens [ ] ones

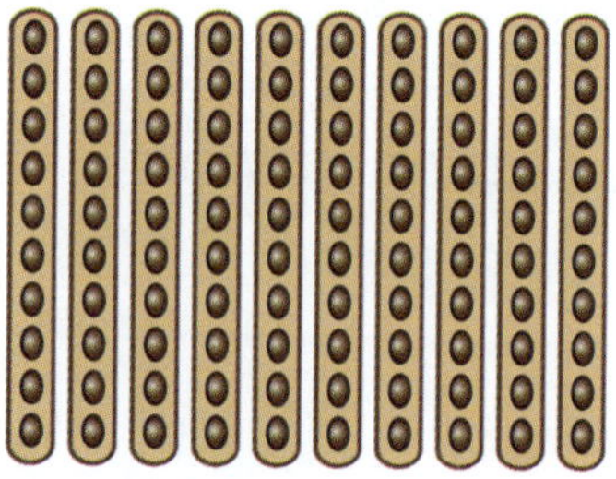

b [ ] hundreds [ ] tens [ ] ones

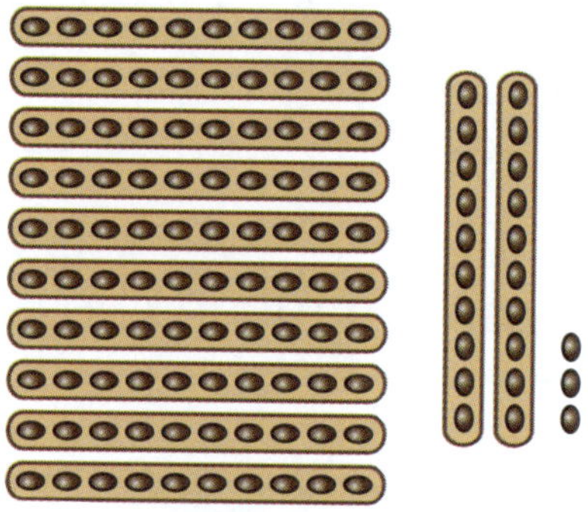

c [ ] hundreds [ ] tens [ ] ones

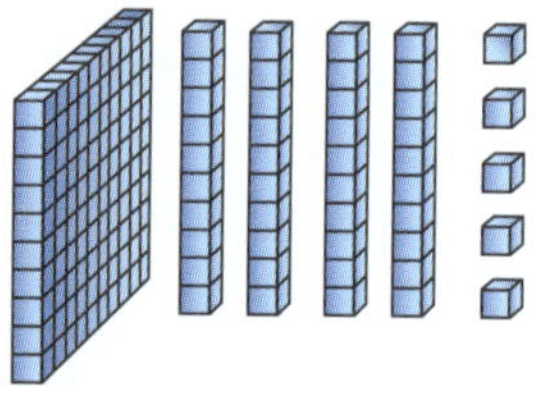

d 1 hundred and ....

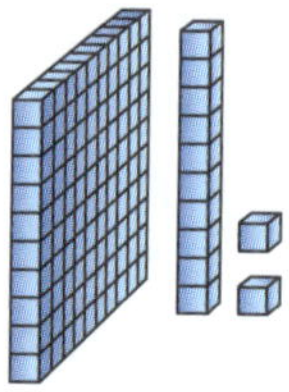

[ ] hundreds [ ] tens [ ] ones

e [ ] hundreds [ ] tens [ ] ones

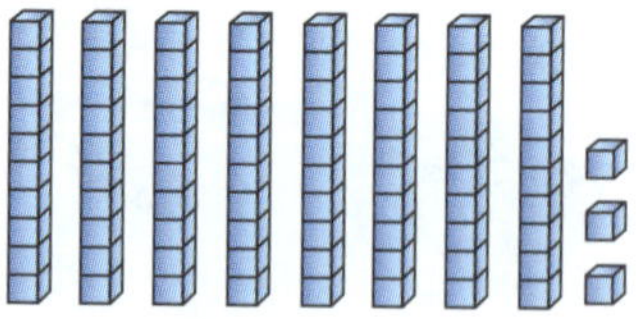

f [ ] hundreds [ ] tens [ ] ones

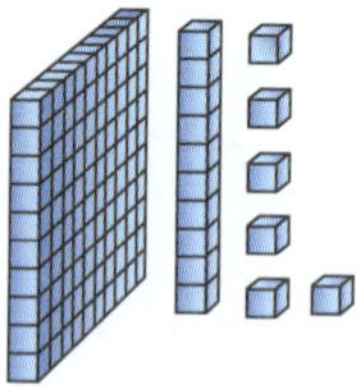

- Use place-value blocks to make these numbers.
  - 106
  - 98
  - 117
  - 123
  - 66
  - 100
- Explain your answers to a friend.

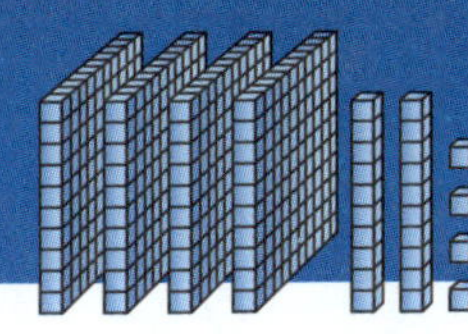

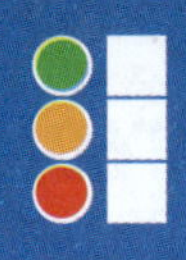

1 Complete each part.

a

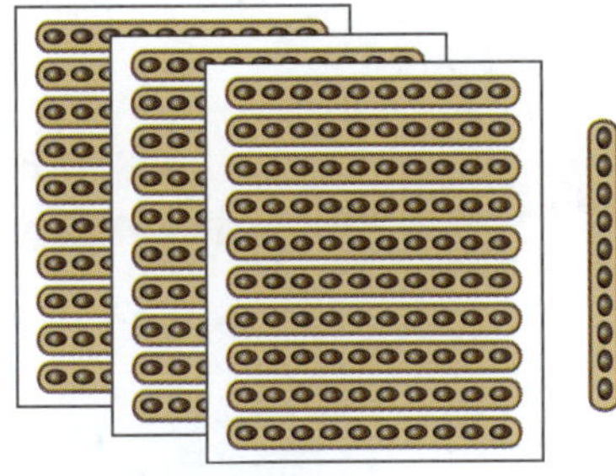
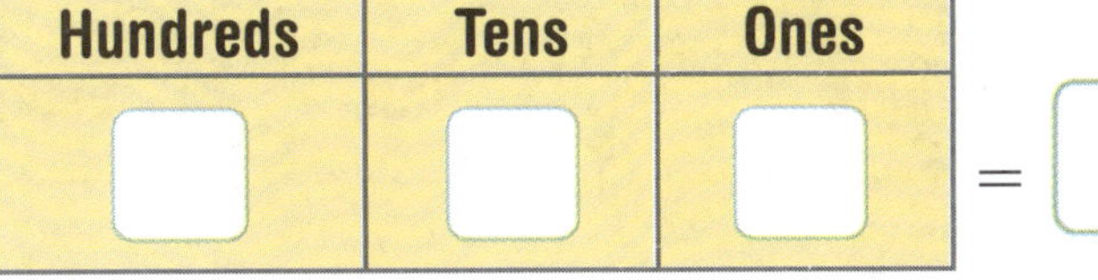

☐ hundreds + ☐ tens + ☐ ones

Hundreds	Tens	Ones
☐	☐	☐

= ☐

b

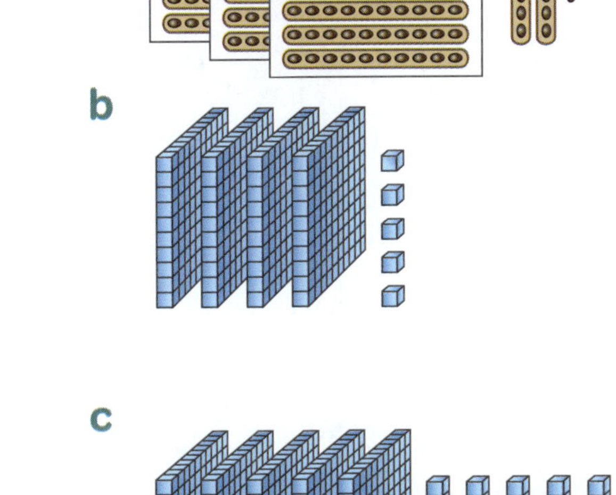

☐ hundreds + ☐ tens + ☐ ones

Hundreds	Tens	Ones
☐	☐	☐

= ☐

c

☐ hundreds + ☐ tens + ☐ ones

Hundreds	Tens	Ones
☐	☐	☐

= ☐

d

☐ hundreds + ☐ tens + ☐ ones

Hundreds	Tens	Ones
☐	☐	☐

= ☐

e

☐ hundreds + ☐ tens + ☐ ones

Hundreds	Tens	Ones
☐	☐	☐

= ☐

- Use place-value blocks to make these numbers.
  - 427
  - 381
  - 836
  - 592
  - 679
  - 764
- Count large groups of objects by grouping in tens and hundreds. Estimate each number first, then check by grouping and counting.

 • *AUSTRALIAN SIGNPOST MATHS NSW 2* • ISBN 9780655709039

16C Informal units of length

ACTIVITY

Use parts of your body to measure objects.
Compare your results with your partner's.

Length	Unit	Estimate	Measurement	Partner's measurement
Width of book	hand span			
Width of book	finger			
Width of window	hand span			

Using a standard informal unit

- Choose a standard length unit (such as an eraser).
- Mark this unit's lengths (1 eraser, 2 erasers, 3 erasers, …) along a ribbon or a piece of string.
- Use your ribbon or string to measure lengths.
- Record your results.

Length	Estimate	Measurement
Length of table	units	units
Width of door	units	units
Width of cupboard	units	units

Would the cupboard fit through the door?

 • *AUSTRALIAN SIGNPOST MATHS NSW 2* • ISBN 9780655709039

Telling the story from data

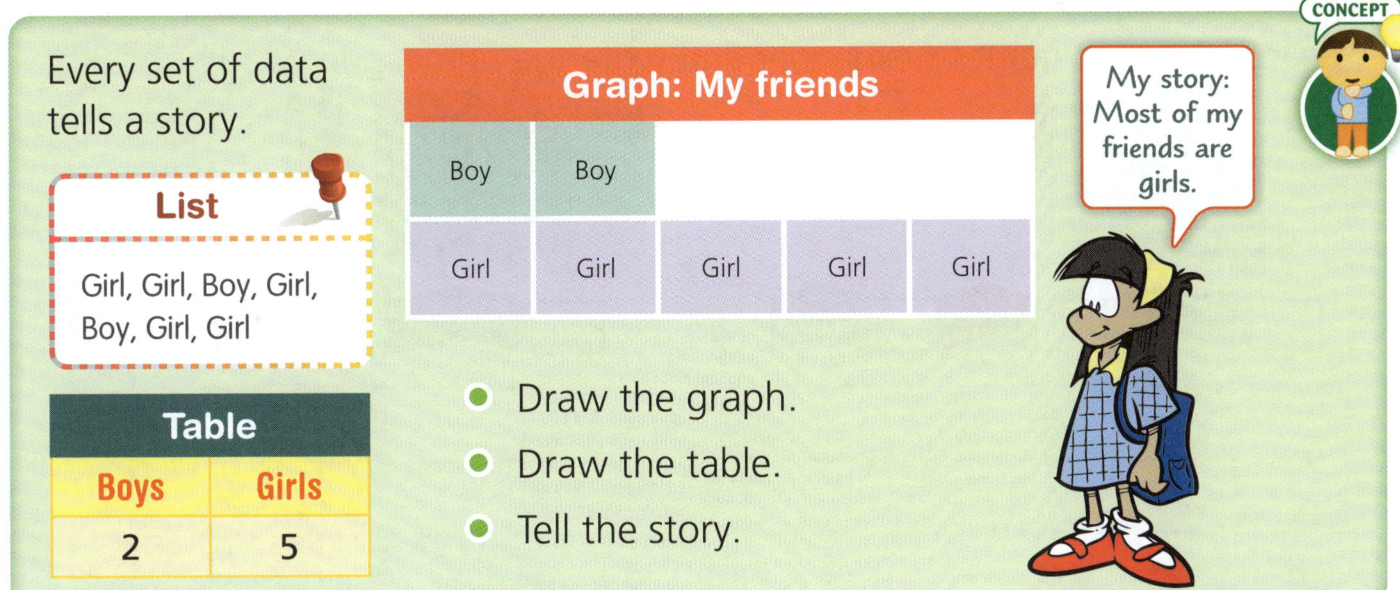

Every set of data tells a story.

List

Girl, Girl, Boy, Girl, Boy, Girl, Girl

Table

Boys	Girls
2	5

Graph: My friends

Boy	Boy			
Girl	Girl	Girl	Girl	Girl

- Draw the graph.
- Draw the table.
- Tell the story.

Complete the graph and table, then tell the story.

1 Children at the party.

List

Boy, Girl, Boy, Boy, Girl, Boy, Boy, Girl

Graph:

At the party

Boys	Girls

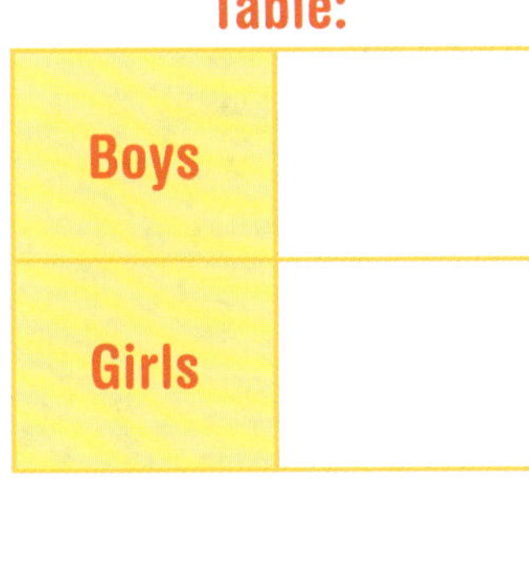

Table:

Boys	
Girls	

Tell the story.

____

____

____

2 Buttons on shirts.

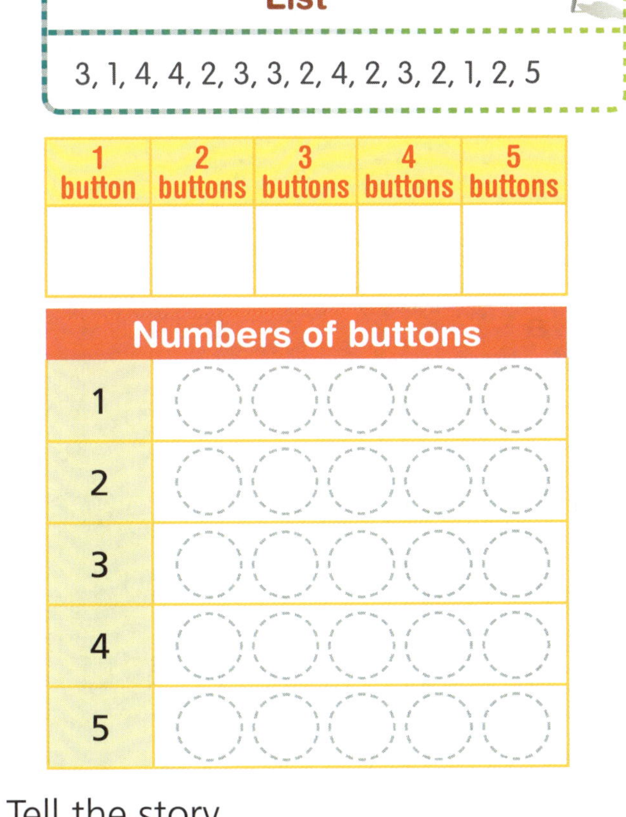

List

3, 1, 4, 4, 2, 3, 3, 2, 4, 2, 3, 2, 1, 2, 5

1 button	2 buttons	3 buttons	4 buttons	5 buttons

Numbers of buttons

1					
2					
3					
4					
5					

Tell the story.

____

____

 • *AUSTRALIAN SIGNPOST MATHS NSW 2* • ISBN 9780655709039

17A Numbers to 1000

1 Complete the numeral expander and write the numeral in words.

a

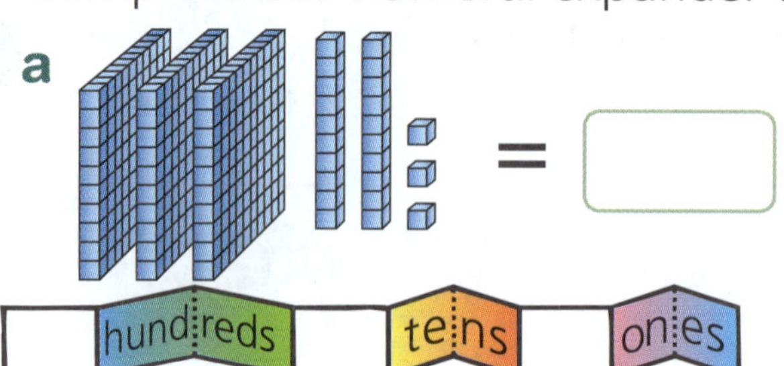

b

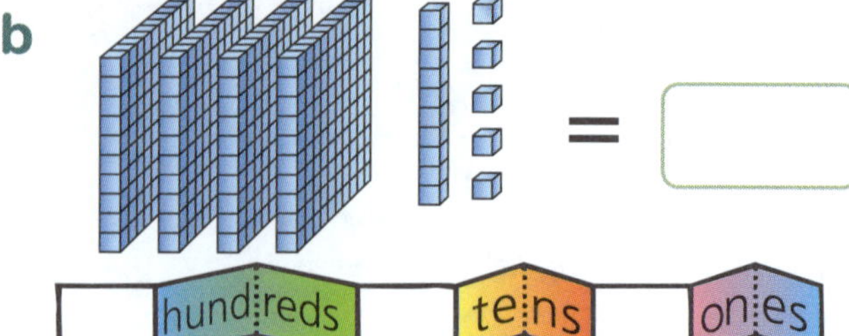

c

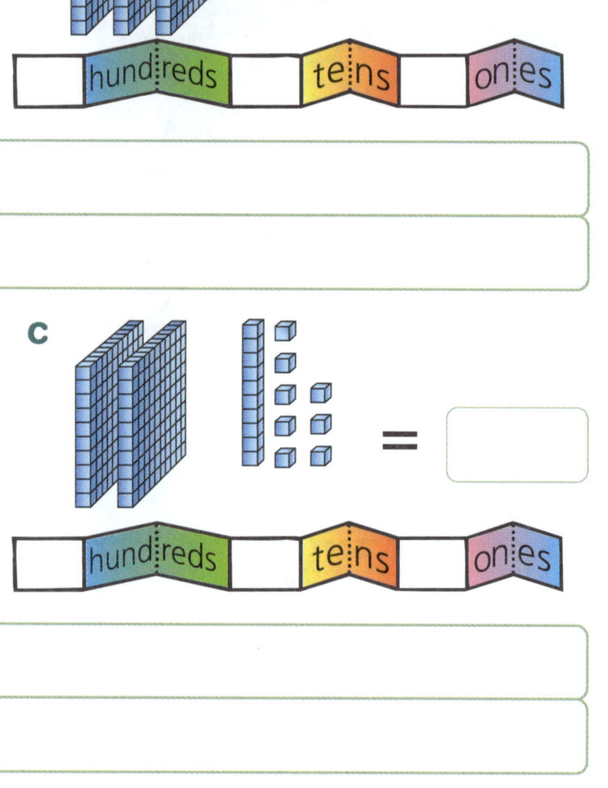

d

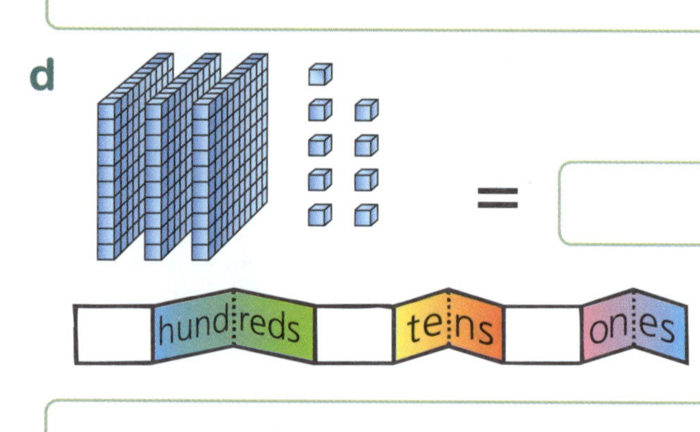

2 Write the next number (the number one more than).

a 213		**b** 407		**c** 728		**d** 699	
e 817		**f** 202		**g** 909		**h** 999	

3 Write the number before (the number one less than).

a 333		**b** 607		**c** 399		**d** 500	
e 206		**f** 891		**g** 468		**h** 1000	

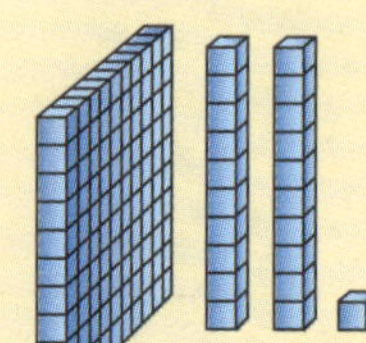

Use place-value blocks to model these numbers.

a 267 **b** 399 **c** 701

d 425 **e** 512 **f** 430

 • *AUSTRALIAN SIGNPOST MATHS NSW 2* • ISBN 9780655709039

Numbers to 1000

CONCEPT

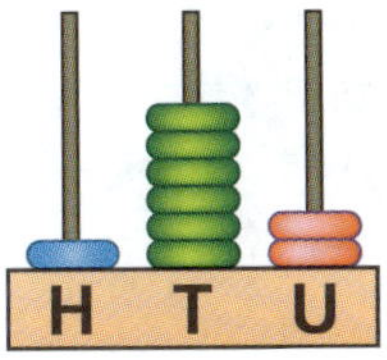

1. Complete each part.

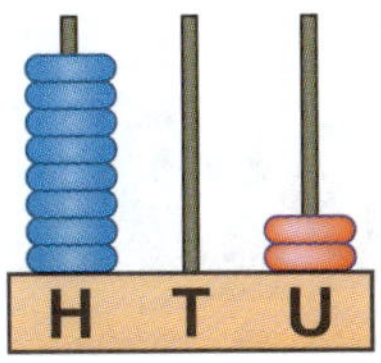

a H T U

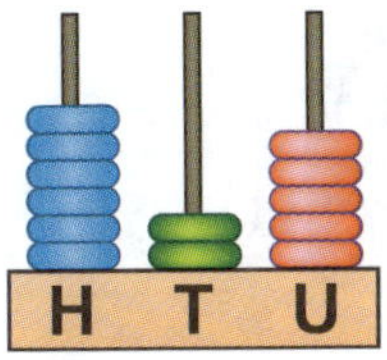

b H T U

c H T U

d H T U

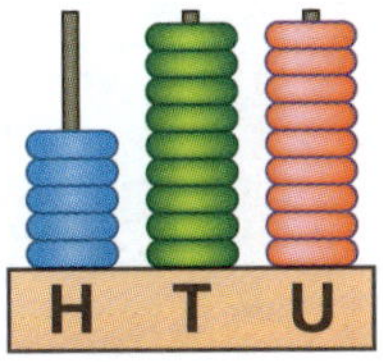

e 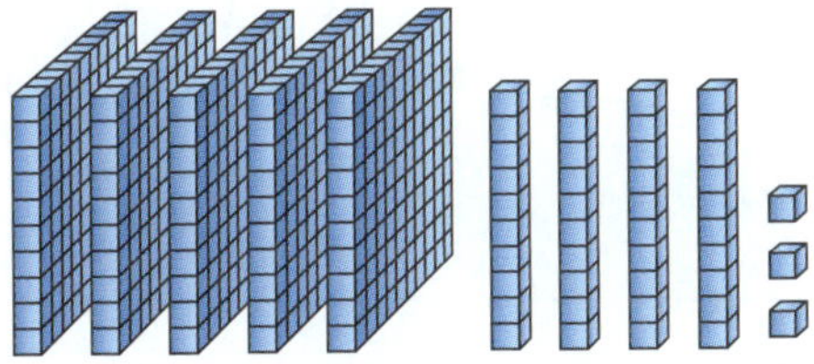 ☐ hundreds + ☐ tens + ☐ ones

Hundreds	Tens	Ones

= ☐

f ☐ hundreds + ☐ tens + ☐ ones

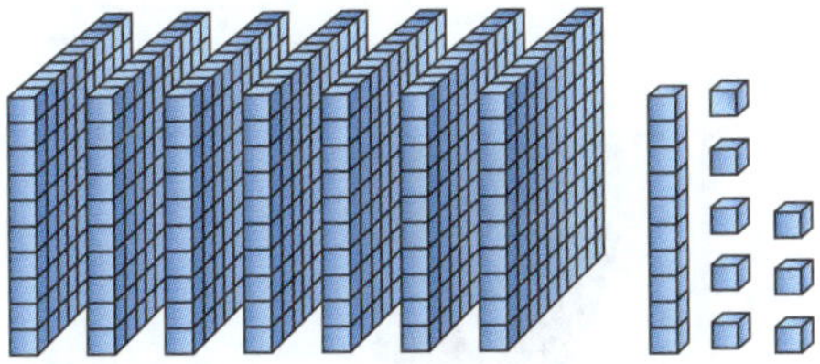

Hundreds	Tens	Ones

= ☐

g ☐ hundreds + ☐ tens + ☐ ones

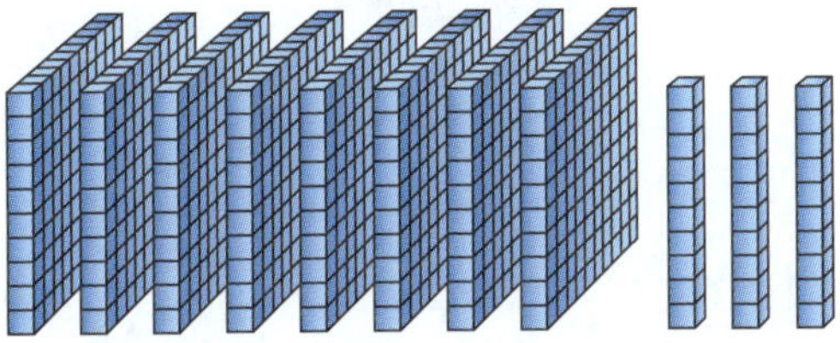

Hundreds	Tens	Ones

= ☐

INVESTIGATION

Use place-value blocks to partition and regroup these numbers. Explain your answer to a friend. For example, 500: 5 hundreds, 50 tens, 500 ones.

- 800
- 300
- 700
- 400
- 265

 • *AUSTRALIAN SIGNPOST MATHS NSW 2* • ISBN 9780655709039

17C Informal units of length

The pencil is 4 ovals long.

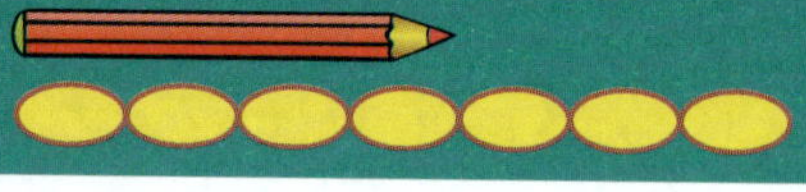

CONCEPT

Is the ones block a good length unit for drawings?

We have a lot of place-value blocks.

We use them with no gaps or overlaps.

We can group them in tens (tens block).

tens block

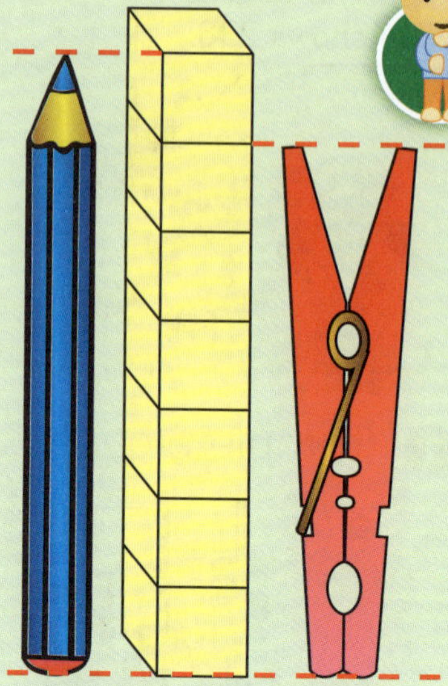

1. Find the lengths of these pictures using the ones block as a unit of length.

a width = ☐ ones blocks height = ☐ ones blocks

b length ☐ ones blocks
height ☐ ones blocks

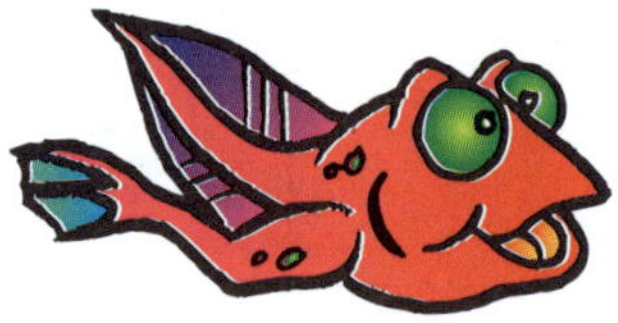

c ? 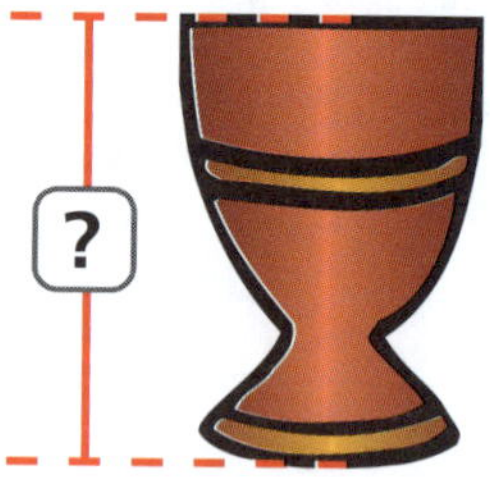
height ☐ ones blocks
width ☐ ones blocks

2. Use place-value blocks to find the length of:

a this page	☐	ones blocks
b your finger	☐	ones blocks
c a pencil	☐	ones blocks
d a pencil case	☐	ones blocks
e a lunch box	☐	ones blocks
f a hand span	☐	ones blocks

Give answers to the nearest ones block.

g Circle the longest object. Underline the shortest object.

17D The metre

This is a very tall bunny.

One metre is about two of your steps.

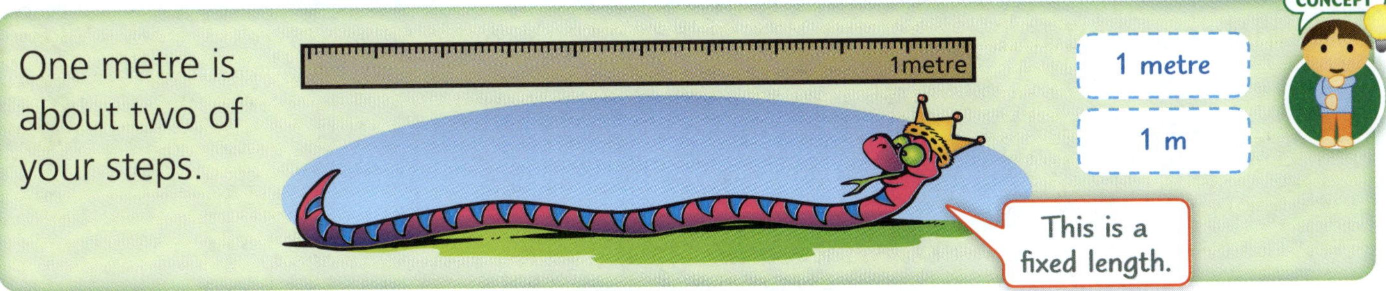

1 Make a list of objects that are about 1 metre long.

2 Cut a strip of paper that is 1 metre long. How many of each item would it take to make 1 m?

Item		Estimate	Measurement
	How many shoes?		
	How many books?		
	How many hands?		

How many of each item can fit into 1 metre?

3 Match each length with the best answer.

- height of a tree
- width of a door
- length of a shoe
- length of a desk

- less than 1 metre
- about 1 metre
- more than 1 metre

Estimate and measure each length to the nearest metre.

Length	Estimate	Measure
length of 10 steps		
length of 20 steps		

Length	Estimate	Measure
classroom length		
classroom width		

18A Numbers to 1000

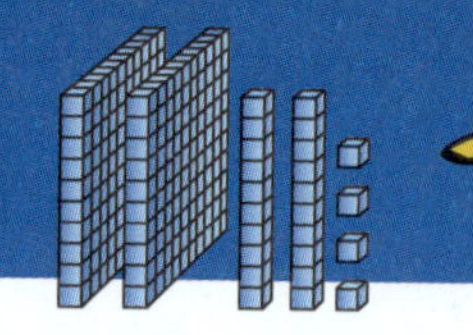

CONCEPT

To make large numbers, count the place-value blocks as you go.

9 hundreds + 2 tens + 5 ones = 925

Hundreds	Tens	Ones
IIIIIIIII	II	IIIII
9	2	5

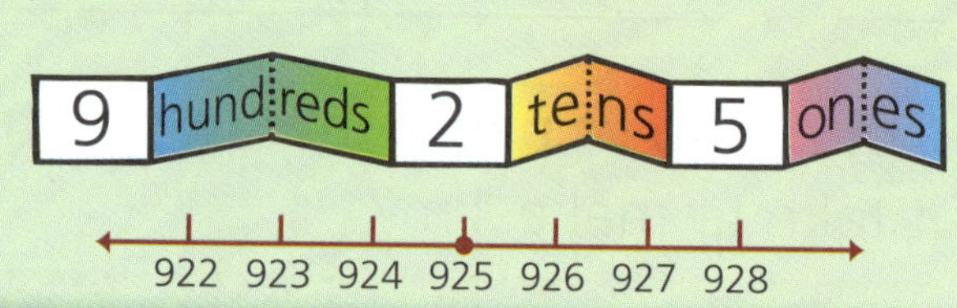

1 Complete each part.

a

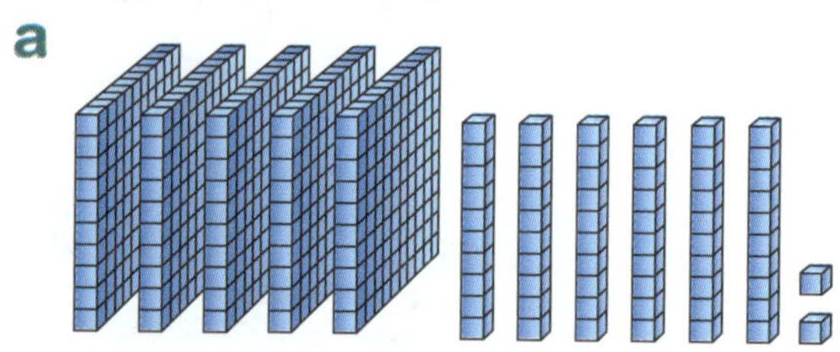

☐ hundreds + ☐ tens + ☐ ones

Hundreds	Tens	Ones
☐	☐	☐

= ☐

b

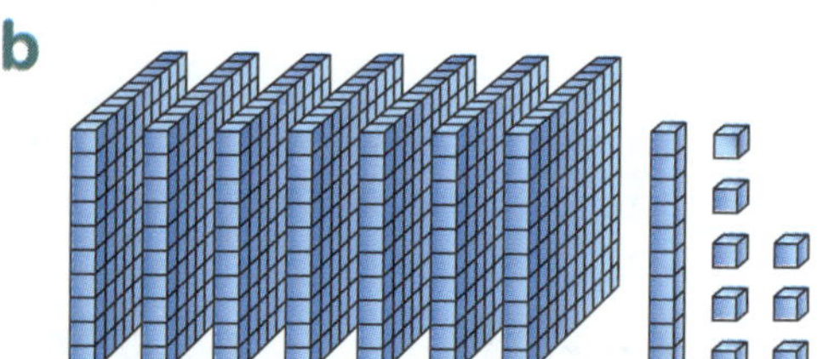

☐ hundreds + ☐ tens + ☐ ones

Hundreds	Tens	Ones
☐	☐	☐

= ☐

c

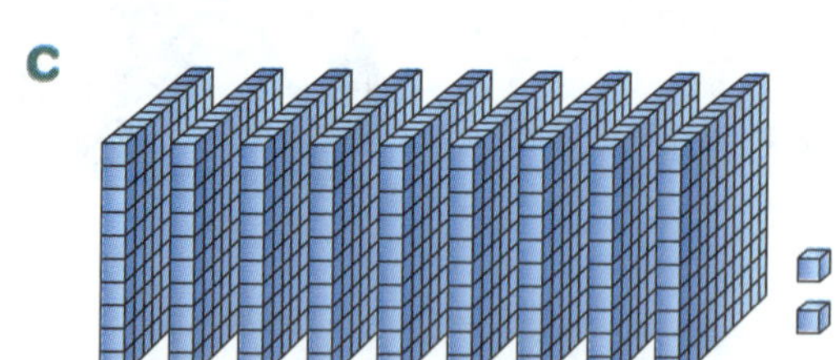

☐ hundreds + ☐ tens + ☐ ones

Hundreds	Tens	Ones
☐	☐	☐

= ☐

ACTIVITY

Model the numbers **703**, **906**, **658**, **812** and **543** using place-value blocks. Write these numbers in order from smallest to largest.

18B Number patterns

If you read down a column, you are saying an "add 10" pattern.

1. On the number chart:

   a count by twos to fifty.

   b count by fives to fifty and circle each number counted.

   c count by tens to one hundred and colour each number counted.

   d What number is before and after:

   ☐ 57 ☐ ?

   ☐ 70 ☐ ?

1	2	3	4	5	6	7	8	9	10
11	12	13	14	15	16	17	18	19	20
21	22	23	24	25	26	27	28	29	30
31	32	33	34	35	36	37	38	39	40
41	42	43	44	45	46	47	48	49	50
51	52	53	54	55	56	57	58	59	60
61	62	63	64	65	66	67	68	69	70
71	72	73	74	75	76	77	78	79	80
81	82	83	84	85	86	87	88	89	90
91	92	93	94	95	96	97	98	99	100

2. Complete the number patterns and describe them.

a

b

c

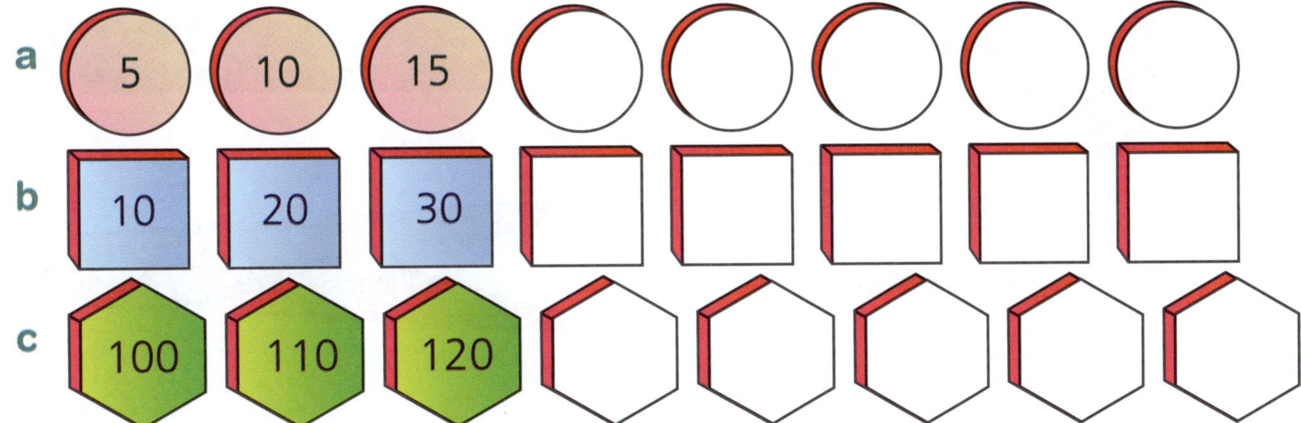

3. Write the **largest** number possible using these digits.

   a 2, 1 ☐ b 6, 9 ☐ c 7, 5 

4. Write the **smallest** number possible using these digits.

   a 1, 2 ☐ b 8, 7 ☐ c 4, 8 ☐

   d 4, 1, 2 ☐ e 6, 9, 1  f 3, 8, 2 

 • *AUSTRALIAN SIGNPOST MATHS NSW 2* • ISBN 9780655709039

18C Centimetres

Use a paper strip or ones blocks to measure in cm.

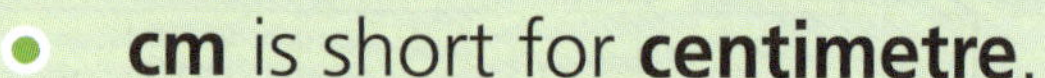

- **cm** is short for **centimetre**.
- 100 centimetres is the same as 1 metre.

100 cm = 1 m

CONCEPT

The grey line is 7 cm long.

1 How long is each line?

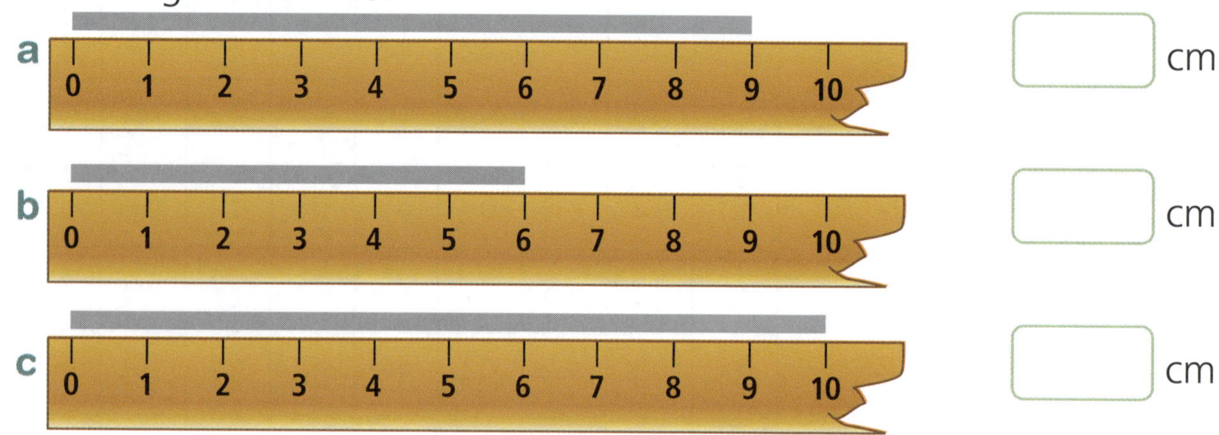

a ☐ cm

b ☐ cm

c ☐ cm

2 Estimate and then measure the length of each item. Write your measurement correct to the nearest centimetre.

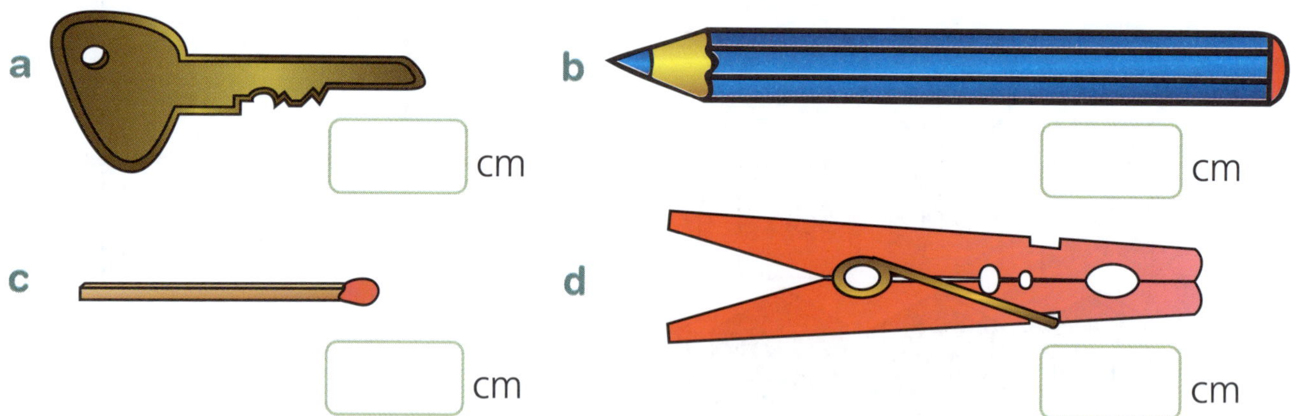

a ☐ cm

b ☐ cm

c ☐ cm

d ☐ cm

3 Estimate and then measure the length of each line. Write your measurement correct to the nearest centimetre.

a ☐ cm

b ☐ cm

c ☐ cm

d ☐ cm

e ☐ cm

 • *AUSTRALIAN SIGNPOST MATHS NSW 2* • ISBN 9780655709039

18D Measuring with centimetres

1 cm is about as long as a ones block.

1 Estimate and then measure the length of each bar. Write the measurements.

2 Estimate and then measure the sides of each shape. Write your measurements.

a cm cm cm cm

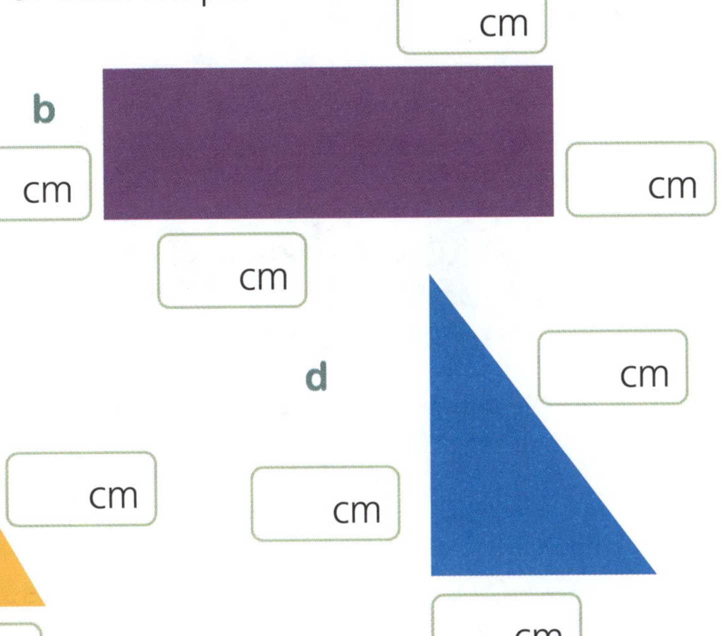

c cm cm cm

e Which of these shapes is the longest all the way around?

Measure items in your classroom to complete the table.

Item	Measurement	Item	Measurement
	Length: cm		Length: cm
	Width: cm		Width: cm
	Height: cm		Height: cm

ACTIVITY

 • *AUSTRALIAN SIGNPOST MATHS NSW 2* • ISBN 9780655709039

19A Number patterns

5, 10, 15, 20 … forwards
8, 6, 4, 2 … backwards

1. Start at 2. Colour every 2nd number green. These are even numbers.

1	2	3	4	5	6	7	8	9	10
11	12	13	14	15	16	17	18	19	20
21	22	23	24	25	26	27	28	29	30

2. Describe the pattern shown on each number line.

a

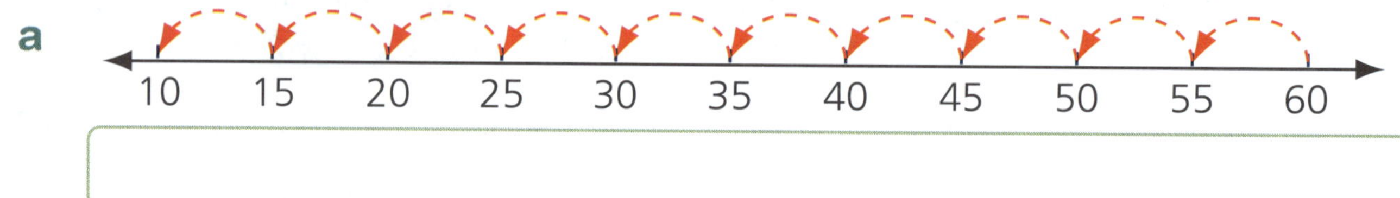

b

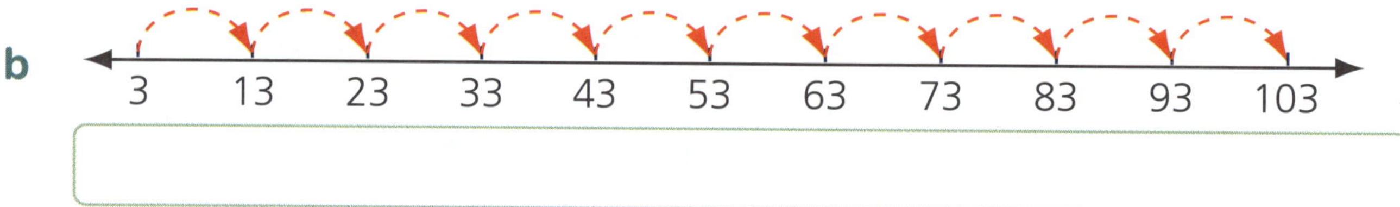

3. Write the missing numbers. Describe the rule for each pattern.

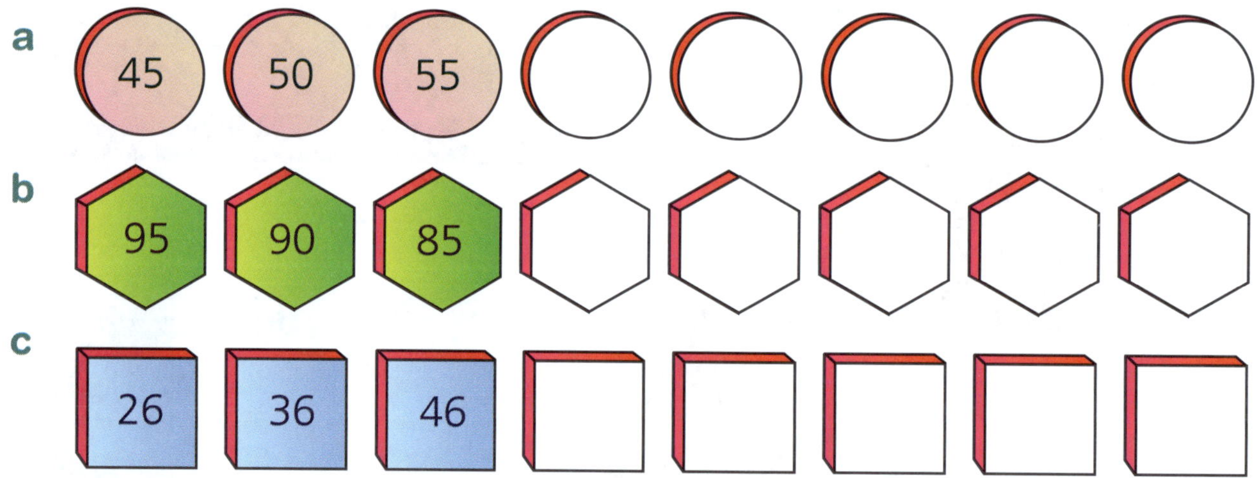

4. Write the next three numbers in each number pattern.

a 64, 54, 44, ______

b 65, 70, 75, ______

c 90, 80, 70, ______

d 37, 47, 57, ______

e 15, 25, 35, ______

f 34, 36, 38, ______

How does a number change as you continue to add ten (56, 66, 76 …)?

Counting by tens

3, 13, 23, 33 ... forwards

96, 86, 76, 66 ... backwards

CONCEPT

Counting numbers to 30

1	2	3	4	5	6	7	8	9	10
11	12	13	14	15	16	17	18	19	20
21	22	23	24	25	26	27	28	29	30

Counting numbers from 100 to 130

101	102	103	104	105	106	107	108	109	110
111	112	113	114	115	116	117	118	119	120
121	122	123	124	125	126	127	128	129	130

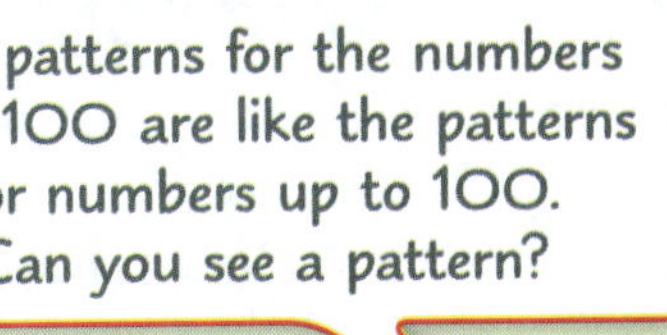

7, 17, 27 ...
107, 117, 127 ...

307, 317, 327, ...
807, 817, 827, ...

1 Complete the tens patterns.

a 7, 17, 27, ___, 47, ___, 67, ___

b 48, 58, 68, ___, ___, ___, ___, ___

c 52, 62, 72, ___, ___, ___, ___, ___

d 39, 49, 59, ___, ___, ___, ___, ___

e 21, 31, 41, ___, ___, ___, ___, ___

2 Complete these counting by tens patterns.

a 103, 113, 123, ___, ___

b 609, 619, 629, ___, ___

c 345, 355, 365, ___, ___

d 432, 442, 452, ___, ___

e 128, 118, 108, ___, ___

f 581, 571, 561, ___, ___

g 345, 355, 365, ___, ___

h 432, 442, 452, ___, ___

i 103, 113, 123, ___, ___

j 609, 619, 629, ___, ___

 • *AUSTRALIAN SIGNPOST MATHS NSW 2* • ISBN 9780655709039

19C Comparing areas

Area is the measure of the amount of surface.

The red triangle has more area than the green triangle.

1 For each pair of shapes, circle the shape that has more area.

a  

b 

c 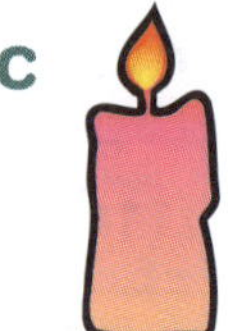 

2 For each group of shapes, circle the largest area and colour the smallest area.

a 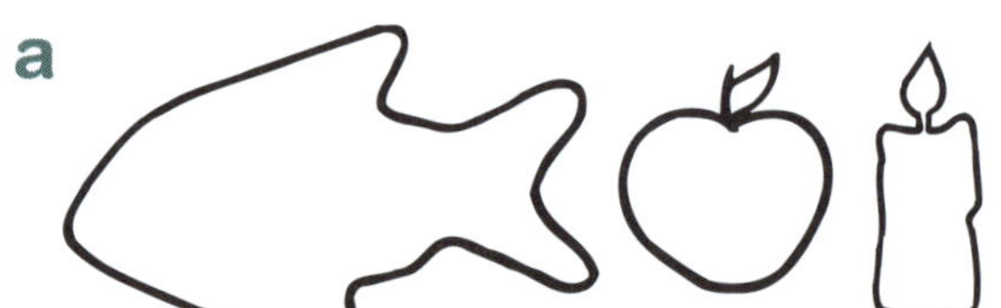

b  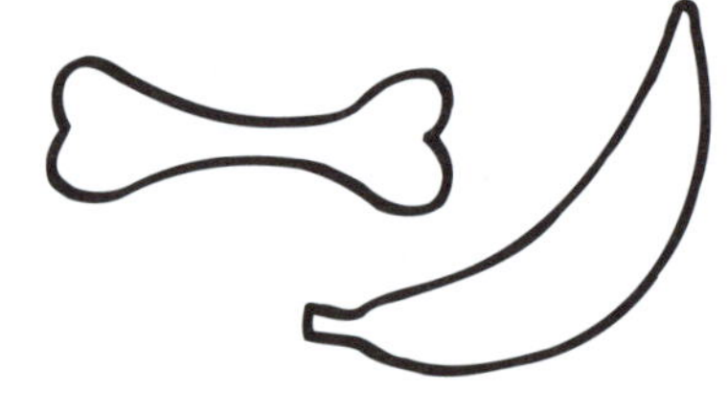

ACTIVITY

Use sheets of newspaper to find the larger area. Count how many sheets it takes to cover one area, then compare this with the next area.

You will need:

- newspaper

- sticky tape

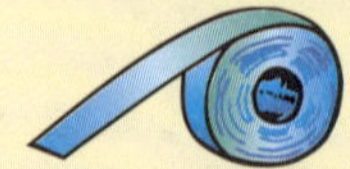

- scissors

First area	Second area	Comparison
Door	Window	
Desk	Gate	
Computer	Television	

Explain the reasoning you used to find your answers.

1 Record the area of these patterns. (They have no gaps or overlaps.)

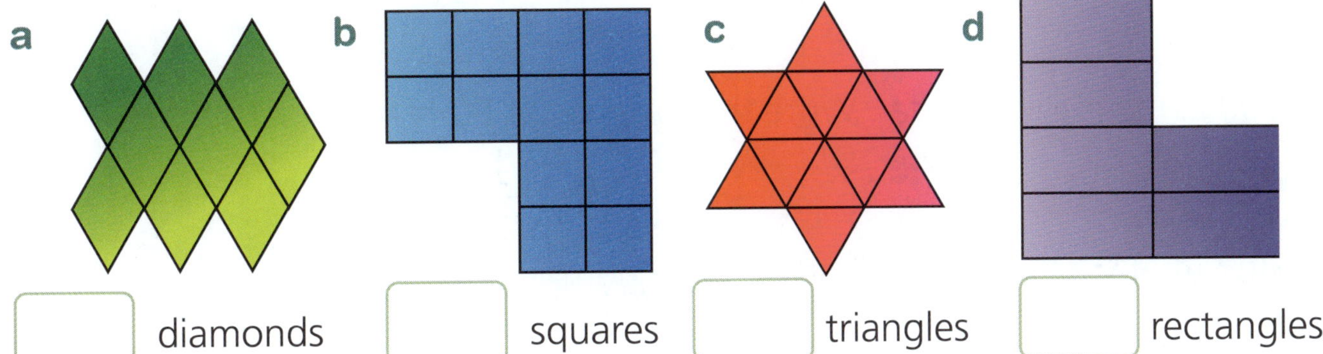

a ☐ diamonds b ☐ squares c ☐ triangles d ☐ rectangles

2 Shape **A** has 3 rows of 3 (or 3 columns of 3). Describe the area of shape:

a **B** ☐

b **C** ☐

Draw a shape with:

c 4 columns of 3. Label it with a **D**.

d 2 rows of 2. Label it with an **E**.

Which shape has:

e the largest area? ☐

f the smallest area? ☐

3 a Estimate which letter has the largest area. How many ones place-value blocks are needed to cover each letter?

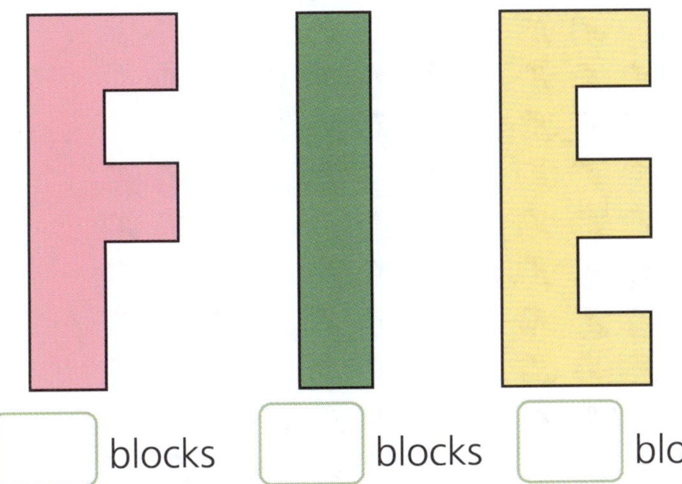

☐ blocks ☐ blocks ☐ blocks

b Which letter has the largest area? ☐

c Which letter has the smallest area? ☐

d Which letter has an area smaller than F? ☐

e Which letter has an area larger than F? ☐

Draw the repeated unit on each shape.

20A Numbers

237 = 2 hundreds, 3 tens and 7 ones
or 23 tens and 7 ones

1 100 cents make a dollar. Write the number of 10 cent coins in:

a $1 ☐ b $2 ☐ c $3 ☐ d $4 ☐

100 cents make a dollar. Write the number of 50 cent coins in:

e $1 ☐ f $2 ☐ g $3 ☐ h $4 ☐

2 How many groups of 100 are in each number?

a 328 ☐ b 740 ☐ c 237 ☐ d 1000 ☐

e 534 ☐ f 198 ☐ g 671 ☐ h 928 ☐

3 Write these numbers as tens and ones (518 = 51 tens and 8 ones).

a 100 = ☐ tens ☐ ones b 135 = ☐ tens ☐ ones

c 456 = ☐ tens ☐ ones d 309 = ☐ tens ☐ ones

e 294 = ☐ tens ☐ ones f 673 = ☐ tens ☐ ones

4 How many groups of 10 are in each number?

a 100 ☐ b 130 ☐ c 456 ☐ d 221 ☐

e 345 ☐ f 510 ☐ g 329 ☐ h 107 ☐

5 Estimate to the nearest 100, the number of marbles that are in this collection. Check by grouping and counting. ☐

Check: ☐

20B Rounding to the nearest ten

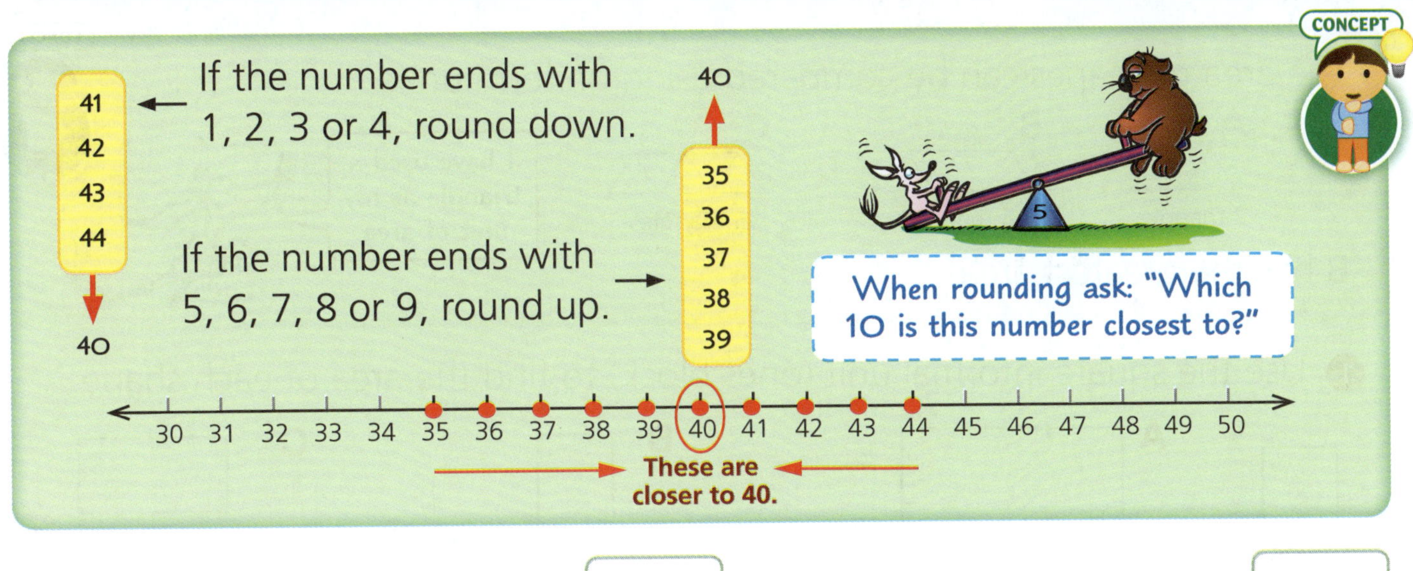

1 a Is 12 closer to 10 or 20? ☐ b Is 77 closer to 70 or 80? ☐

c Is 21 closer to 20 or 30? ☐ d Is 89 closer to 80 or 90? ☐

e Is 33 closer to 30 or 40? ☐ f Is 106 closer to 100 or 110? ☐

2 Round to the nearest 10.

a 24 ☐ b 36 ☐ c 63 ☐ d 47 ☐

e 92 ☐ f 74 ☐ g 59 ☐ h 81 ☐

i 103 ☐ j 115 ☐ k 123 ☐ l 127 ☐

3 a Circle the numbers that round to 60.

57 58 67 68 64

61 69 62 65 52 56

Digits below 5 are rounded down.

b Circle the numbers that round to 90.

99 87 86 97 94 83

93 85 89 88 91

4 Write true (T) or false (F) for each statement.

a 124 is less than 130. ☐ b 127 is more than 117. ☐

c 111 is more than 101. ☐ d 60 is less than 47. ☐

e 35 is more than 76. ☐ f 93 is less than 100. ☐

 • *AUSTRALIAN SIGNPOST MATHS NSW 2* • ISBN 9780655709039

20C Area using informal units

CONCEPT

The area of shapes can be compared using uniform units.

A — 3 triangles

B — 6 triangles

C — 5 triangles

B has the greatest area.

I have used a triangle as my unit of area.

1 Use the square informal unit (ones block) to find the area of each shape.

1 unit of area

A ☐ units

B ☐ units

C ☐ units

D ☐ units

E ☐ units

F ☐ units

G ☐ units

a Which shape has the smallest area? ☐

b Which shape has the largest area? ☐

c Which shape has the same area as F? ☐

d Which shape has the same area as E? ☐

e Which shape has double the area of D? ☐

2 Draw a shape with an area of 8 square units on this grid.

ACTIVITY

Use ones blocks as a uniform unit to compare three different areas.

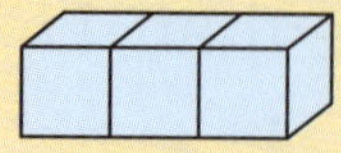

Record the comparison of these areas using drawings, numerals and units.

20D Area of a rectangle

CONCEPT

Join the marks on opposite sides of this rectangle to find how many squares will cover the area.

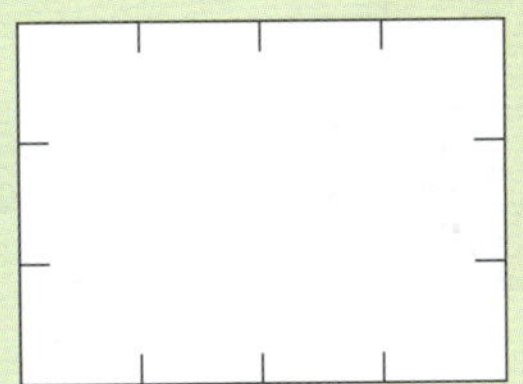

The area of the rectangle = ☐ square units

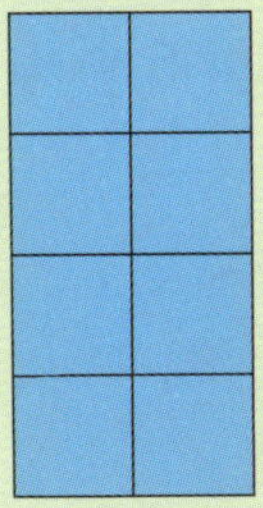

Here are 4 rows of 2 squares.

Area = 2 + 2 + 2 + 2
= 8 square units

1 Find the area of each rectangle using the small squares as a unit of area.

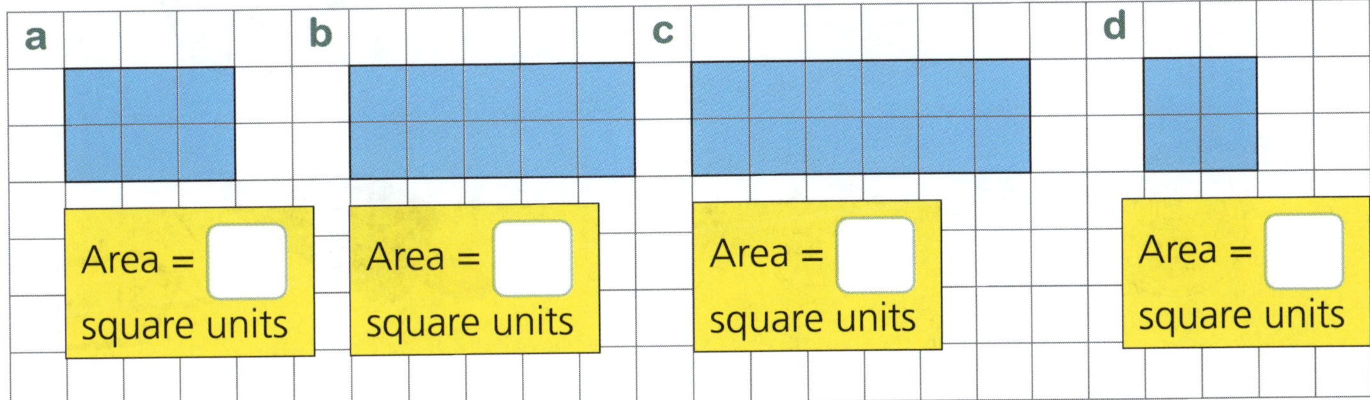

a Area = ☐ square units

b Area = ☐ square units

c Area = ☐ square units

d Area = ☐ square units

2 Use a pencil and ruler to draw the grid on each rectangle and find the area.

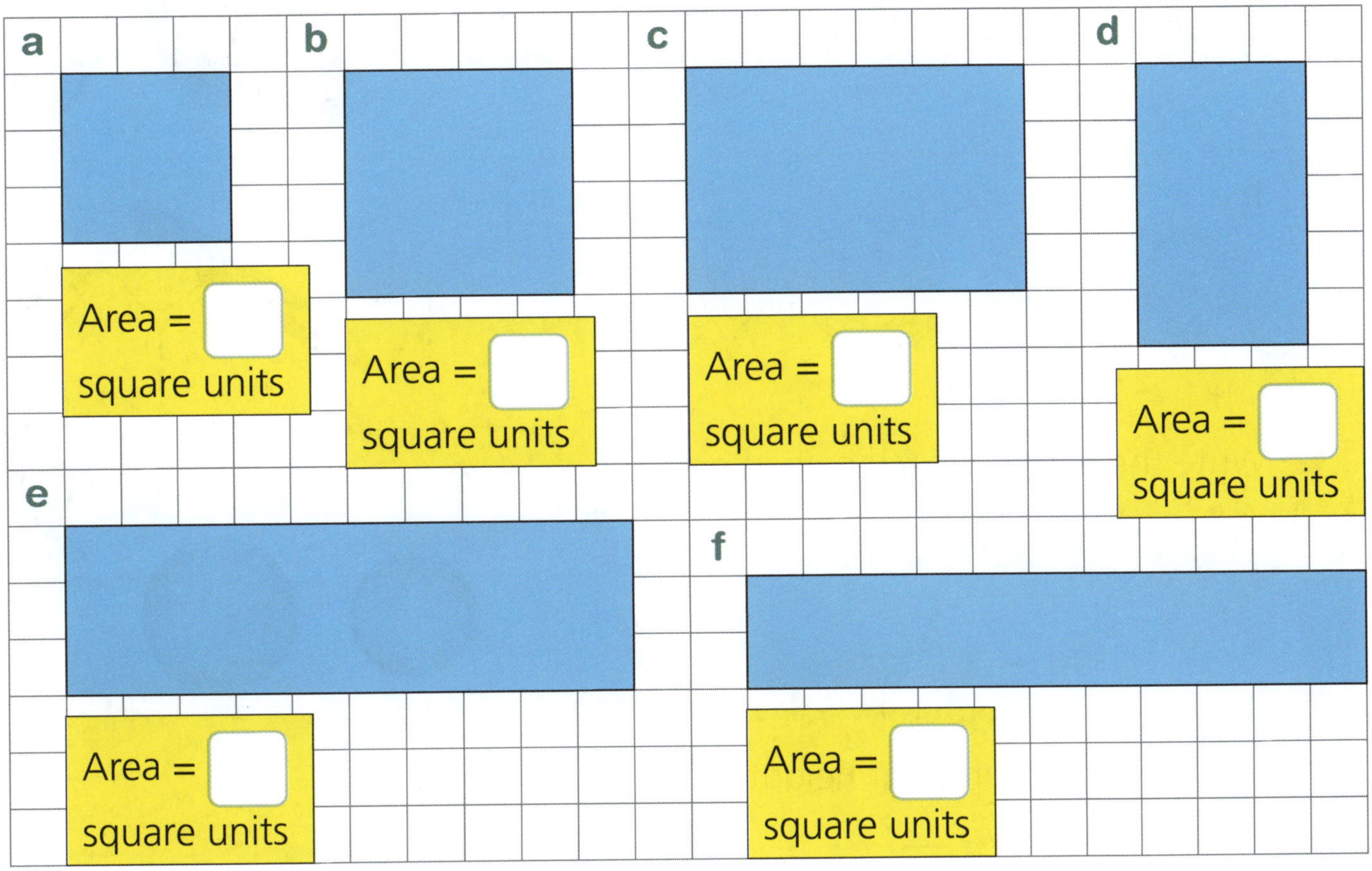

a Area = ☐ square units

b Area = ☐ square units

c Area = ☐ square units

d Area = ☐ square units

e Area = ☐ square units

f Area = ☐ square units

21A Value of coins

$1 = 100c

1 Colour the coins you would use to pay for each item.

a

b

c

d

e

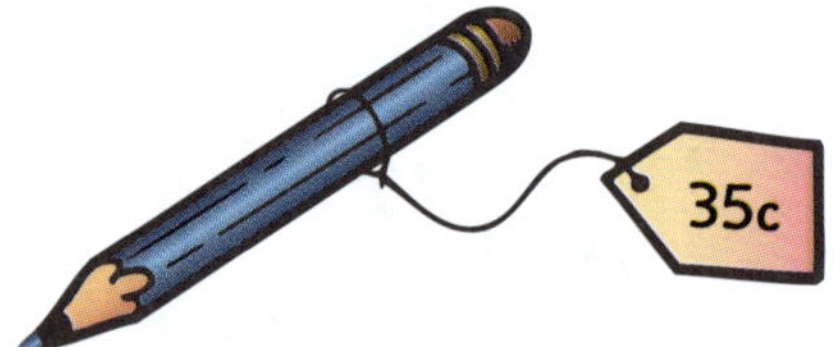

2 Write the **value** of each coin in order.

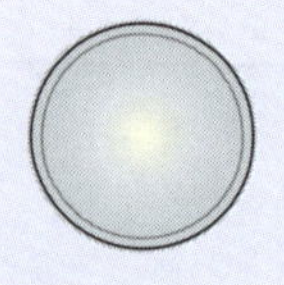
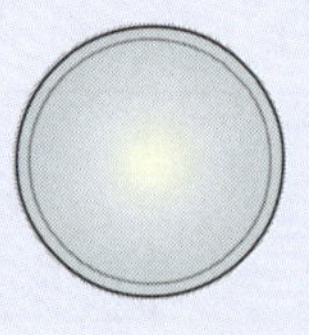
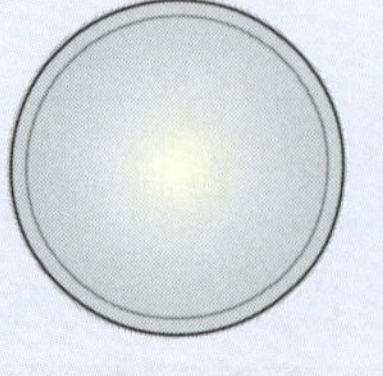
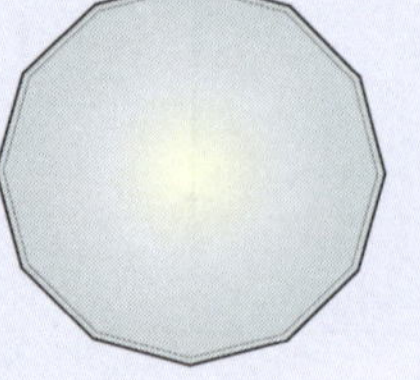
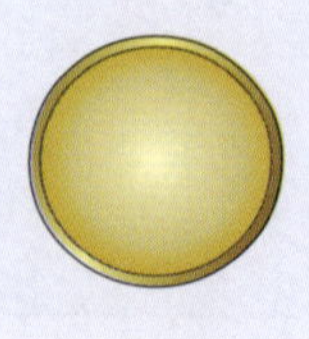
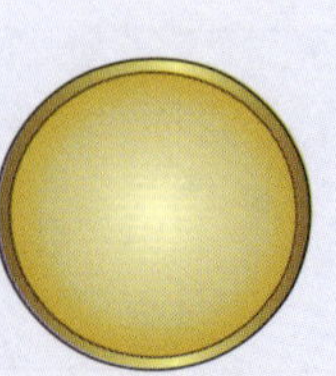

List the coins in order of height.

CONCEPT

To find "how much", start with the highest value and count on.

$1 ... $1.50 ... $1.70 ... $1.80

1 How much money?

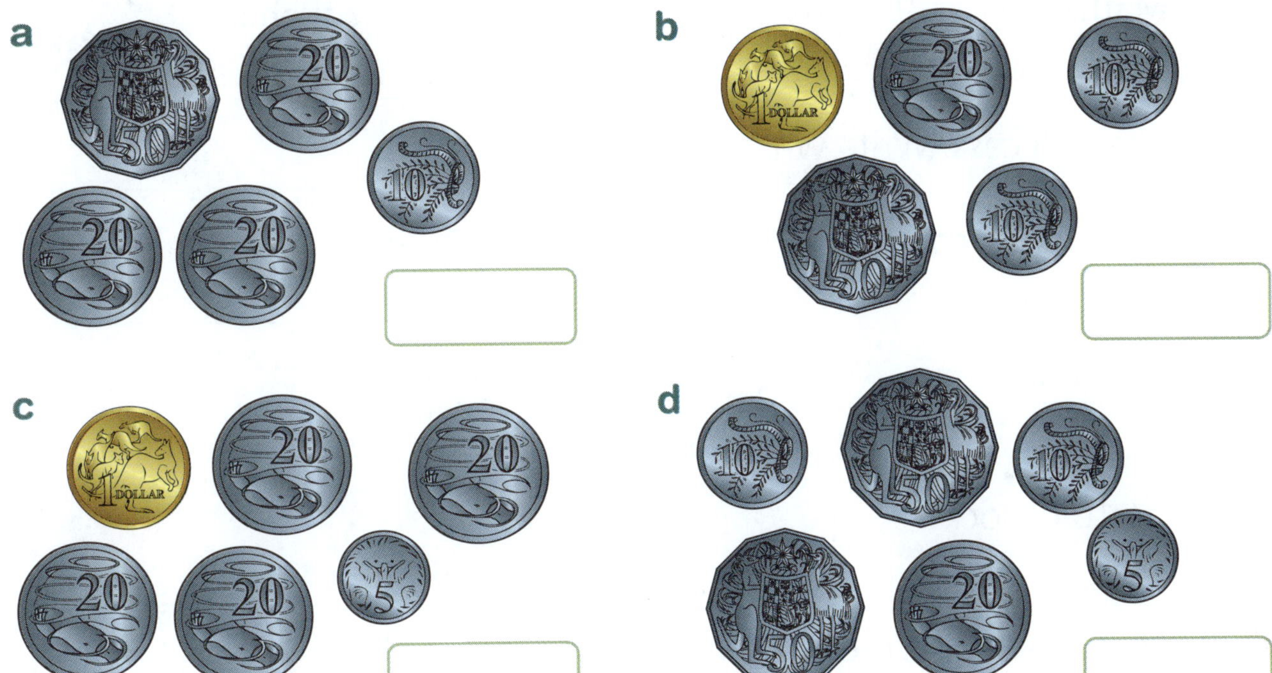

2 Colour the coins you would use to pay for each item.

21C Duration using hours

1 Discuss this timetable.

7 o'clock	8 o'clock	9 o'clock	10 o'clock	11 o'clock
Wake up	Breakfast	School starts	Reading groups	Recess
12 o'clock	**1 o'clock**	**2 o'clock**	**3 o'clock**	**4 o'clock**
Maths	Lunch	Sport	School ends	Swimming lesson

2 It is now Recess. Write what happened:

a four hours ago

b two hours ago

c at 10 o'clock

3 It is now Recess. Write what will happen:

a two hours from now

b four hours from now

c at 4 o'clock

Make up a daily timetable of your own.
Talk about it with a friend.

21D Duration using weeks

1 Discuss this simple calendar.

Week 1	Week 2	Week 3	Week 4	Week 5
Start of term	Sam's party	Start project	Zoo excursion	Test
Week 6	**Week 7**	**Week 8**	**Week 9**	**Week 10**
Class talks	Drama night	Project due	Parent–teacher interviews	End of term

2 It is now Week 5. Write what happened:

a four weeks ago

b two weeks ago

c last week

3 It is now Week 5. Write what will happen:

a next week

b three weeks from now

c four weeks from now

INVESTIGATION

Use a real calendar to find the number of weeks or days to or after events.

22A Amounts to $2

1 How much money is in each moneybox?

a

b

CONCEPT

$1 + 20c + 5c = $1.25

To find "how much", start with the highest coin and count on.

c

d

e

2 What is the value of each set of coins?

a

b

c

d

INVESTIGATION

- Use coins or play money to make 50c in as many ways as you can. My number of ways = ☐.
- Use coins to make amounts chosen by your partner.

 • *AUSTRALIAN SIGNPOST MATHS NSW 2* • ISBN 9780655709039

22B Value of coins

1 Colour the coins you would use to pay for each item.

a

b

c

d

2 Use skip counting to find each total.

a          cents

b         cents

3 Write the value of each group of coins.

a

b

c

 • *AUSTRALIAN SIGNPOST MATHS NSW 2* • ISBN 9780655709039

22C Prisms and cylinders

1 Write **prism** or **cylinder** under each object.

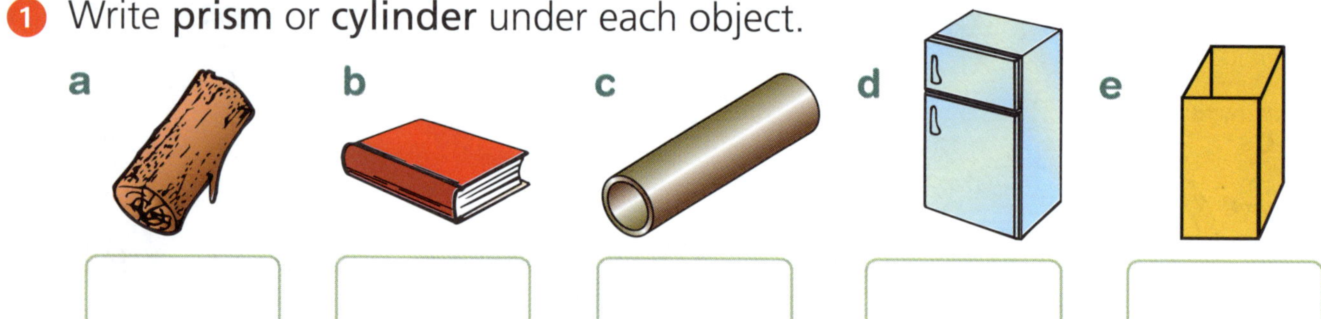

2 Use the words **flat surface**, **curved surface**, **slides** and **rolls** to describe each object.

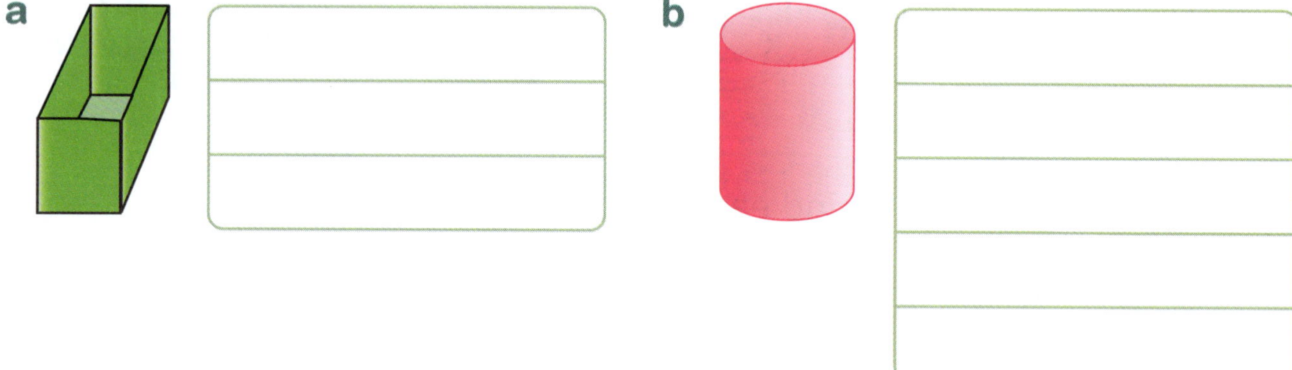

3 Write the names of the flat surfaces on each object.

Use plasticine or playdough to make models of prisms and cylinders.

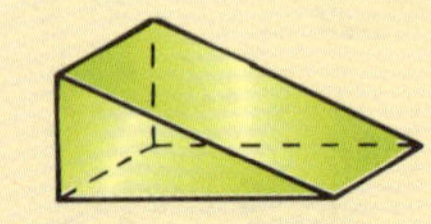

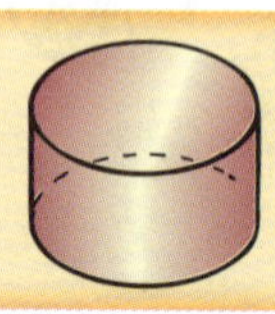

22D 3D objects

1 Match each item with one of the 3D objects. Write the name of each 3D object below it.

2 Draw a picture of the 3D object. How many surfaces does it have?

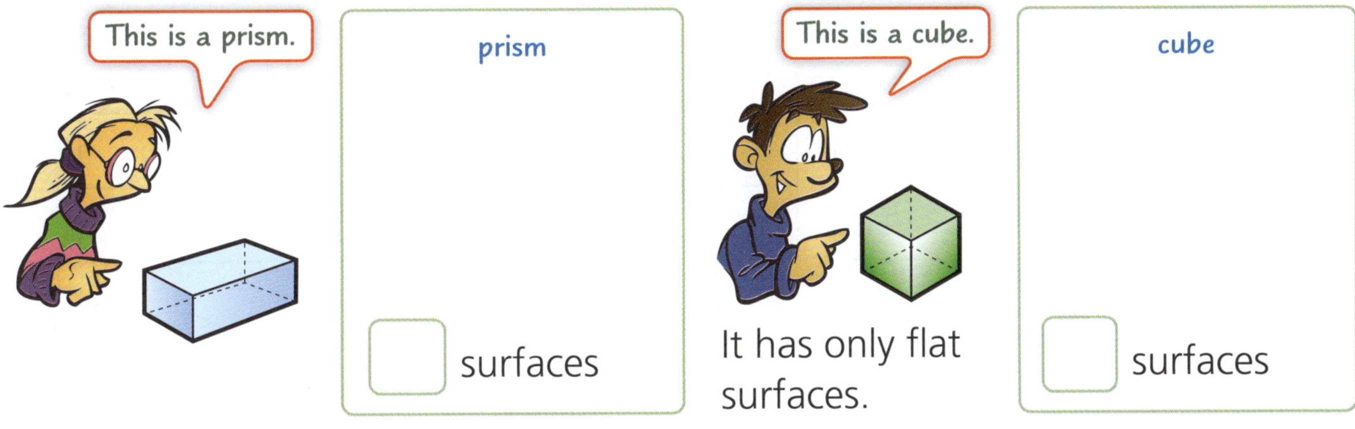

3 Complete the pattern of cylinders and cones.

INVESTIGATION

Use plasticine or playdough to make a model of a:

- sphere
- cylinder
- cone
- prism

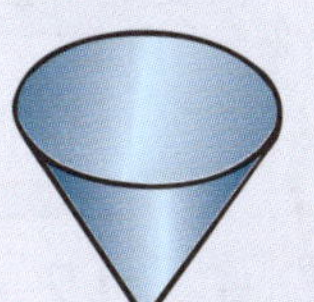

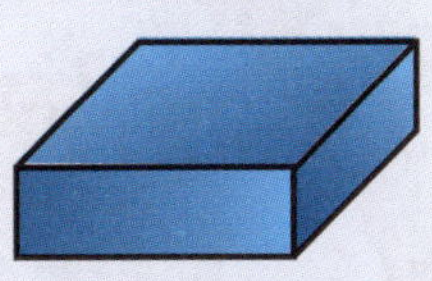

Build a model of a cube using toothpicks. Use BluTack, plasticine or playdough to hold the parts together.

  ISBN 9780655709039

23A Building to the next 10

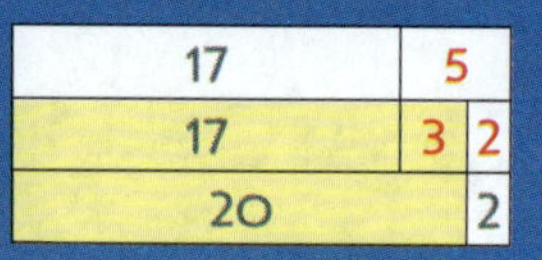

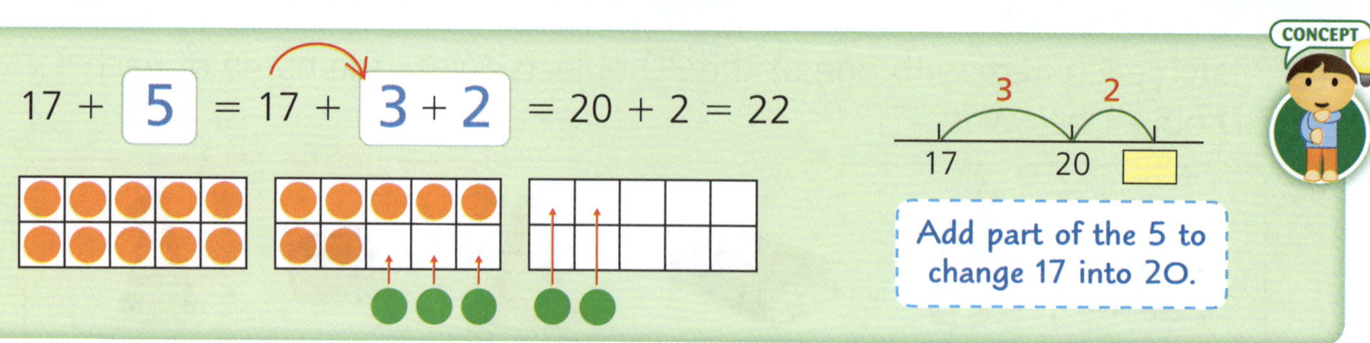

1. Use the ten frames to build to the next 10.

a 18 + 4 = 18 + ☐ + ☐ = 20 + ☐ = ☐

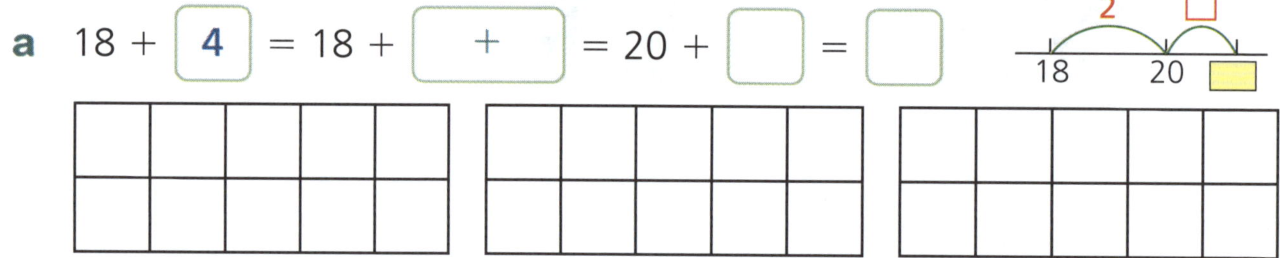

b 17 + 6 = 17 + ☐ + ☐ = 20 + ☐ = ☐

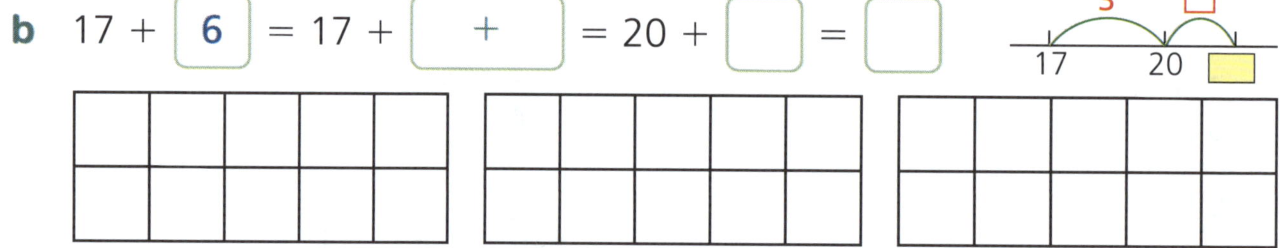

c 19 + 5 = 19 + ☐ + ☐ = 20 + ☐ = ☐

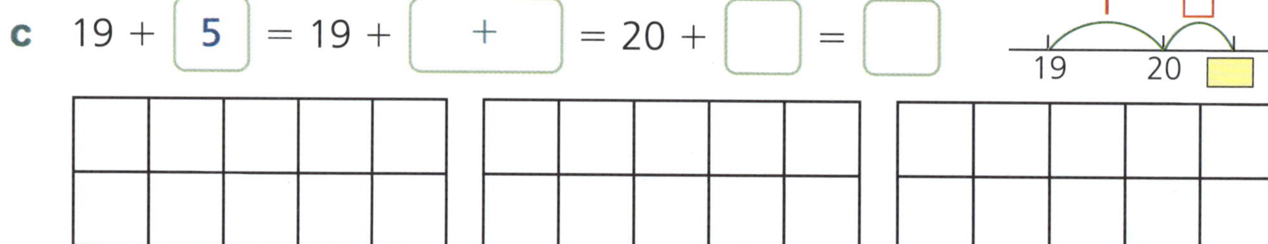

Use ten frames to make up questions of your own.

2. Help Banjo find the answers using the "build to the next 10" strategy.

 • *AUSTRALIAN SIGNPOST MATHS NSW 2* • ISBN 9780655709039

23B Building to the next 10

1 a $37 + 6 = 37 + 3 + \square$
$= 40 + \square$
$= \square$

b $25 + 7 = 25 + 5 + \square$
$= 30 + \square$
$= \square$

c $44 + 8 = 44 + 6 + \square$
$= 50 + \square$
$= \square$

d $73 + 9 = 73 + 7 + \square$
$= 80 + \square$
$= \square$

2 a $56 + 5 = 56 + 4 + \square$
$= \square + \square$
$= \square$

b $69 + 4 = 69 + 1 + \square$
$= \square + \square$
$= \square$

c $38 + 6 = 38 + 2 + \square$
$= \square + \square$
$= \square$

d $84 + 8 = 84 + 6 + \square$
$= \square + \square$
$= \square$

3 Try to do these in your head.

a $38 + 3 = \square$

b $54 + 7 = \square$

c $77 + 8 = \square$

d $29 + 5 = \square$

e $46 + 9 = \square$

f $55 + 7 = \square$

g $35 + 5 = \square$

h $89 + 4 = \square$

Ask: What is needed to reach the next 10?

23C Volume

CONCEPT

Volume is the amount of space within a container or taken up by a model. To measure volume, we use blocks of equal size. These are placed so that there are no gaps.

Capacity is the amount a container can hold. Usually this is a liquid.

1 Estimate which has the largest volume, then count the blocks.

A

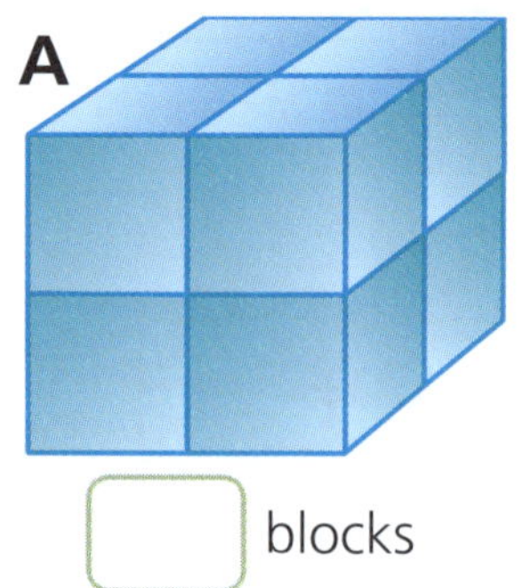

☐ blocks

B

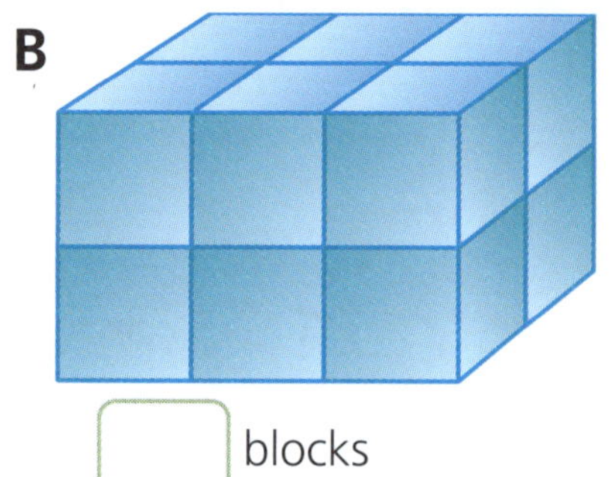

☐ blocks

C

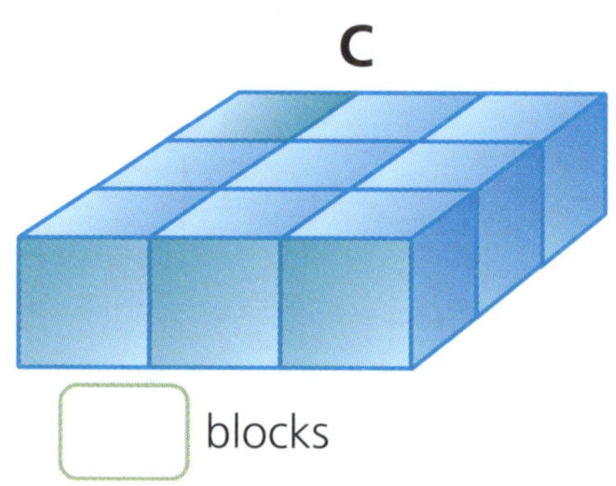

☐ blocks

Order the volumes of these models (smallest to largest). ☐

2 Estimate which has the smallest volume, then count the blocks.

D

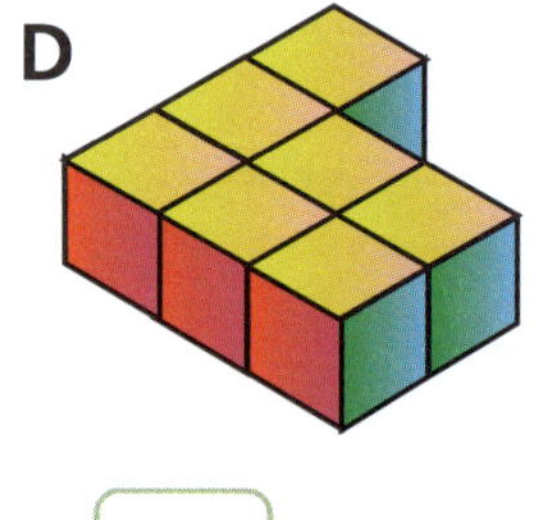

☐ blocks

E

☐ blocks

F

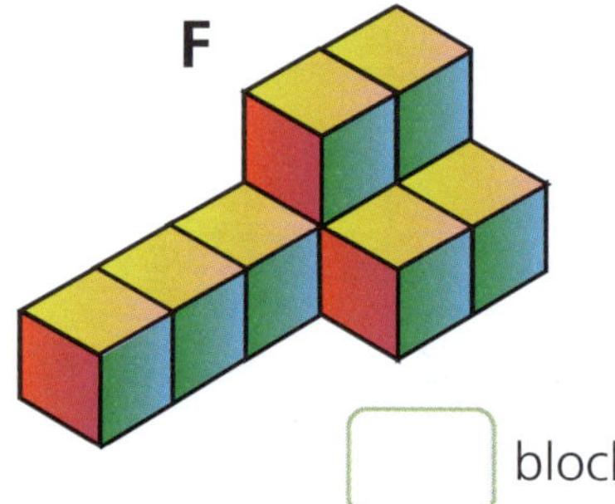

☐ blocks

Order the volumes of these models (smallest to largest). ☐

ACTIVITY

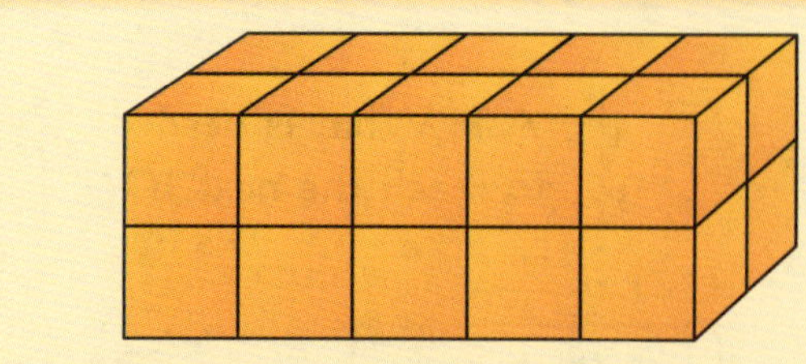

- Use ones blocks to make 3 models.
  Compare and order them according to their volumes.
- Record the results using drawing (or photos), numerals and words.

 ISBN 9780655709039

23D Comparing volume

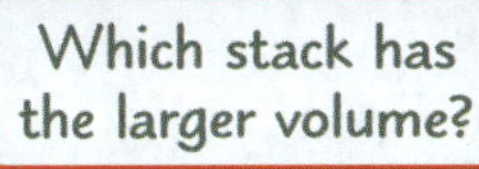

1 a How many blocks in A? ☐ b How many blocks in B? ☐

c Which model contains more blocks? ☐ d Which model has the larger volume? ☐

e Which model has the smaller volume? ☐

2 Count the number of blocks in each model.

a  ☐

b 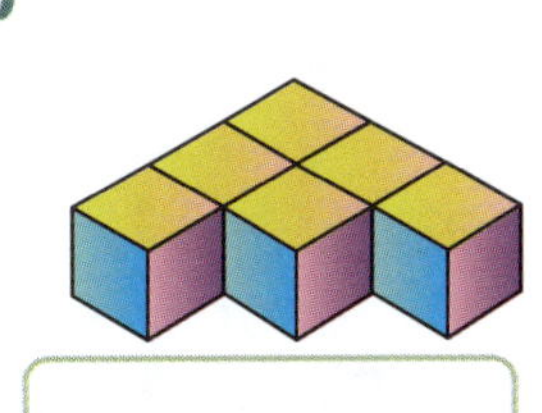 ☐

c  ☐

d  ☐

e  ☐

f 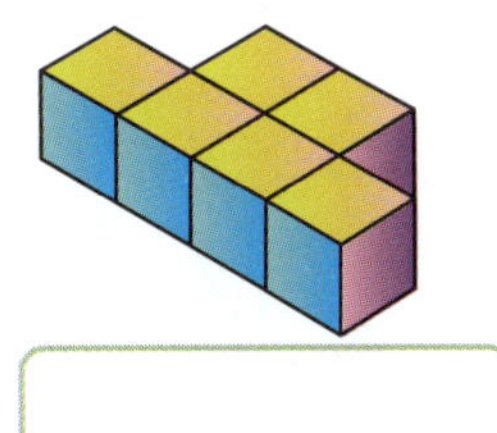 ☐

g  ☐

h  ☐

i  ☐

3 a Draw lines to join models that have the same volume.
b Circle the model with the greatest volume.
c Tick the model with the least volume.

Pack ones blocks into 3 small boxes so there are no gaps. Order the 3 boxes from smallest volume to largest volume.

24A Split strategy (addition)

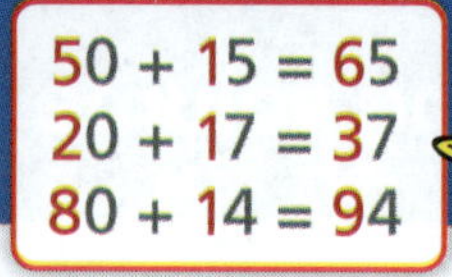

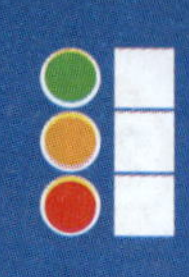

CONCEPT

32 = 30 + 2

Add the tens, add the ones.

= (30 + 40) + (2 + 5) Add tens.

= 77 Add ones.

This is the split strategy.

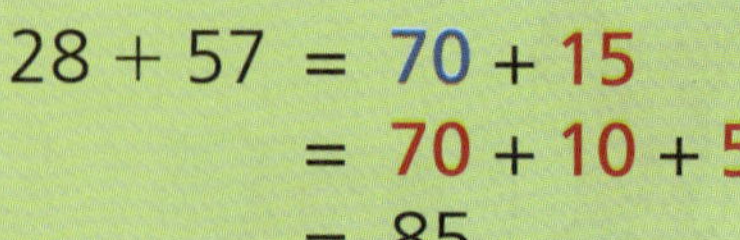

28 + 57 = 70 + 15
= 70 + 10 + 5
= 85

1 Use the split strategy to find the answers.

a 23 + 15
= (20 + 10) + (3 + 5)
= ☐

b 32 + 47
= (30 + 40) + (2 + 7)
= ☐

c 16 + 71
= (10 + 70) + (6 + 1)
= ☐

d 52 + 24
= ☐ + ☐
= ☐

e 81 + 15
= ☐ + ☐
= ☐

f 45 + 54
= ☐ + ☐
= ☐

2 Use the split strategy. (Remember that 70 + 11 = 81.)

a 57 + 24 = 50 + 20 + 7 + 4
= 70 + 11
= ☐

b 38 + 59 = 30 + 50 + 8 + 9
= 80 + 17
= ☐

c 28 + 25 = ☐ + ☐ + ☐ + ☐
= ☐
= ☐

d 46 + 38 = ☐ + ☐ + ☐ + ☐
= ☐
= ☐

3 Try to do these in your head. If you find it hard, use your own paper.

a 32 + 64 = ☐

b 33 + 44 = ☐

c 74 + 16 = ☐

d 52 + 28 = ☐

e 17 + 37 = ☐

f 38 + 16 = ☐

24B Split strategy (addition)

1 **a** 48 + 6 = 40 + 8 + 6
= 40 + 14
= 40 + 10 + 4
= ______

b 56 + 9 = 50 + 6 + 9
= 50 + 15
= 50 + 10 + 5
= ______

c 35 + 16 = 30 + 5 + 10 + 6
= (30 + 10) + (5 + 6)
= 40 + 11
= ______

d 16 + 57 = 10 + 6 + 50 + 7
= (10 + 50) + (6 + 7)
= 60 + 13
= ______

2 **a** 16 + 67 = (10 + 60) + (6 + 7)
= ______ + ______
= ______

b 39 + 35 = (30 + 30) + (9 + 5)
= ______ + ______
= ______

c 64 + 26 = (60 + 20) + (4 + 6)
= ______ + ______
= ______

d 52 + 29 = (50 + 20) + (2 + 9)
= ______ + ______
= ______

3 **a** 43 + 29 = ______ + ______
= ______

b 37 + 56 = ______ + ______
= ______

c 74 + 26 = ______ + ______
= ______

d 42 + 39 = ______ + ______
= ______

4 **a** 44 + 38 = ______

b 57 + 17 = ______

Can you do it in your head?

24C Ordering masses

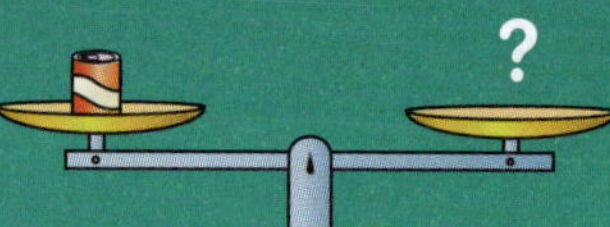

1 **a** Handle (heft) three objects like these and estimate which one is the lightest and which one is the heaviest.

b Order the objects from lightest to heaviest.

______ (lightest), ______, ______ (heaviest)

c Use blocks and a balance scale to find the mass of each object.

Object	Mass
	blocks
	blocks
	blocks

d Write a sentence about the masses of the three objects.

______

______

2 Choose other sets of three objects and compare them in the same way. Record your results.

Object	Mass
	marbles
	marbles
	marbles

Object	Mass

24D Balance scales

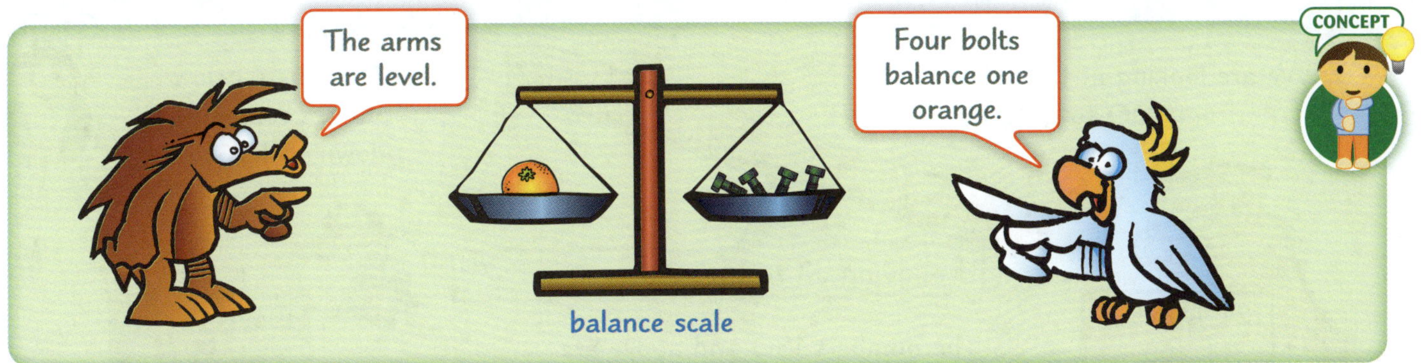

ACTIVITY

1 Use a balance scale to find answers to these.

a 2 marbles balance ☐ craft sticks.

b 1 ball balances ☐ marbles.

c 8 pens balance ☐ marbles.

d 4 pens balance ☐ craft sticks.

e 1 tennis ball balances ☐ marbles.

f 5 marbles balance ☐ blocks.

2 a Choose a unit to use and balance some plasticine. Change the shape of the plasticine and weigh it again. What did you find?

plasticine

plasticine

?

b Choose a different unit to weigh the plasticine. Do you think you will need more of these units? Weigh the plasticine. What did you discover?

 • *AUSTRALIAN SIGNPOST MATHS NSW 2* • ISBN 9780655709039

25A Rounding to the nearest 100

Closer to 100. Closer to 200

100 150 200

1 **a** Is 195 closer to 100 or 200? ☐ **b** Is 387 closer to 300 or 400? ☐

c Is 218 closer to 200 or 300? ☐ **d** Is 427 closer to 400 or 500? ☐

e Is 647 closer to 600 or 700? ☐ **f** Is 566 closer to 500 or 600? ☐

g Is 753 closer to 700 or 800? ☐ **h** Is 341 closer to 300 or 400? ☐

2 Round these numbers to the nearest hundred.

a 135	☐	**b** 370	☐	**c** 523	☐	**d** 458	☐
e 240	☐	**f** 761	☐	**g** 895	☐	**h** 616	☐
i 694	☐	**j** 407	☐	**k** 249	☐	**l** 583	☐

3 **a** Circle the numbers that round to 200.

140 207 124 180 168

105 196 243 318 251 149

b Circle the numbers that round to 500.

423 521 442 509 604

454 598 495 583 531 450

If the 10s column is 5 or more, round up.

4 Write true (T) or false (F) for each statement.

a 286 is more than 267. ☐ **b** 152 is less than 125. ☐

c 347 is less than 352. ☐ **d** 450 is more than 419. ☐

e 728 is less than 782. ☐ **f** 932 is more than 975. ☐

 • *AUSTRALIAN SIGNPOST MATHS NSW 2* • ISBN 9780655709039

25B Building to the next 10

Multiples of 10 end in zero.
10, 20, 30 …
or 80, 130, 250, 870 …

To add a one-digit number, use part of it to reach the next ten, then add on the other part.

4 more are needed to get from 116 to 120.

How many are needed to get from 108 to 110? ☐

101	102	103	104	105	106	107	108	109	110
111	112	113	114	115	116	117	118	119	120
121	122	123	124	125	126	127	128	129	130

How many are needed to get from 126 to 130? ☐

a 17 + 8
= (17 + 3) + 5
= 20 + 5
= 25

b 79 + 5
= (79 + 1) + 4
= 80 + 4
= 84

c 385 + 9
= (385 + 5) + 4
= 390 + ☐
= ☐

d 273 + 8
= (273 + 7) + 1
= 280 + ☐
= ☐

1

71	72	73	74	75	76	77	78	79	80

How many more do you need to get to 80 if you start from:

a 78? ☐ **b** 76? ☐ **c** 79? ☐ **d** 75? ☐ **e** 77? ☐

2

271	272	273	274	275	276	277	278	279	280

How many more do you need to get to 280 if you start from:

a 279? ☐ **b** 277? ☐ **c** 275? ☐ **d** 274? ☐ **e** 278? ☐

3

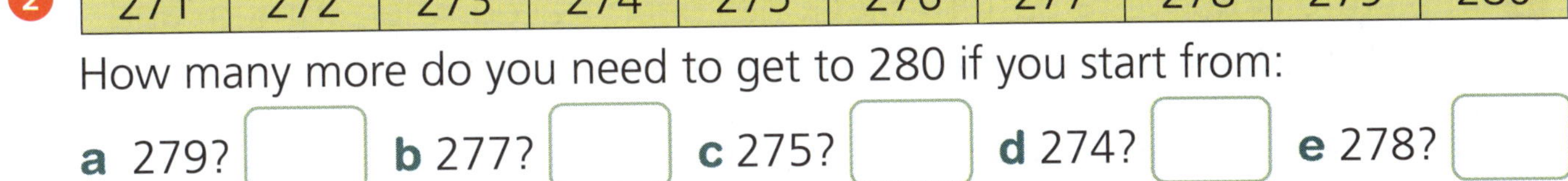

How many more do you need to reach 150 if you start from:

a 146? ☐ **b** 147? ☐ **c** 148? ☐ **d** 143? ☐ **e** 145? ☐

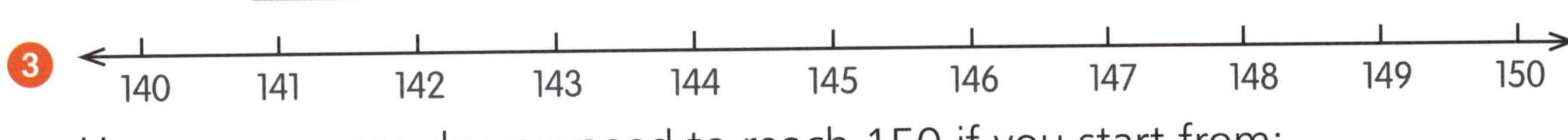

4 257 + 8 = (257 + 3) + 5 = 260 + 5 = 265

Use this method to answer these questions.

a 278 + 4 ☐ **b** 279 + 3 ☐ **c** 277 + 5 ☐

d 274 + 8 ☐ **e** 277 + 7 ☐ **f** 279 + 6 ☐

g 186 + 7 ☐ **h** 349 + 6 ☐ **i** 418 + 9 ☐

 • *AUSTRALIAN SIGNPOST MATHS NSW 2* • ISBN 9780655709039

25C Turning a shape

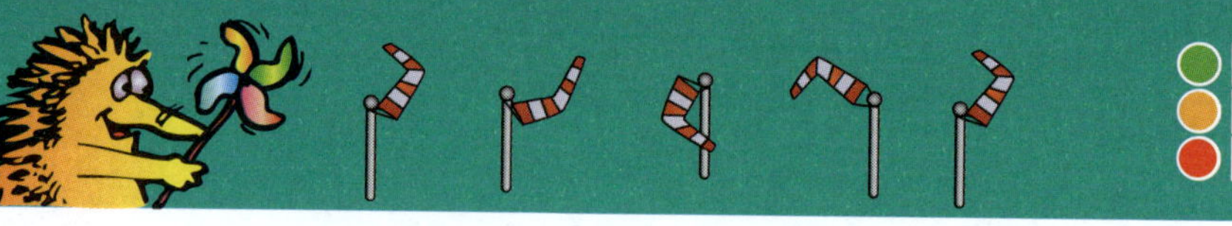

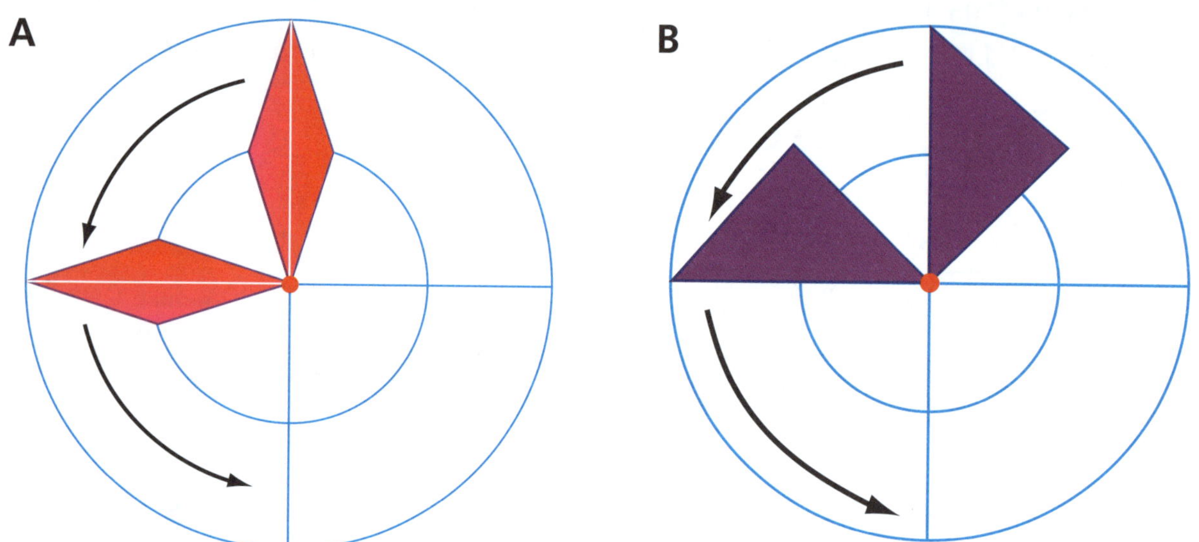

1. In picture **A** above, a pattern block has been traced, then turned, and then traced and turned again. Complete the pattern.

   a Does the shape change as it turns?

   b Does the shape turn around a point?

2. Complete pattern **B**.

ACTIVITY

- Play **Simon Says**, giving directions using the words "slide" and "turn".
- Write two sets of directions.

Slide 2 steps forward.
Slide 4 steps back.

Turn to the right.

Turn to the left.

 • *AUSTRALIAN SIGNPOST MATHS NSW 2* • ISBN 9780655709039

25D Turning shapes

1

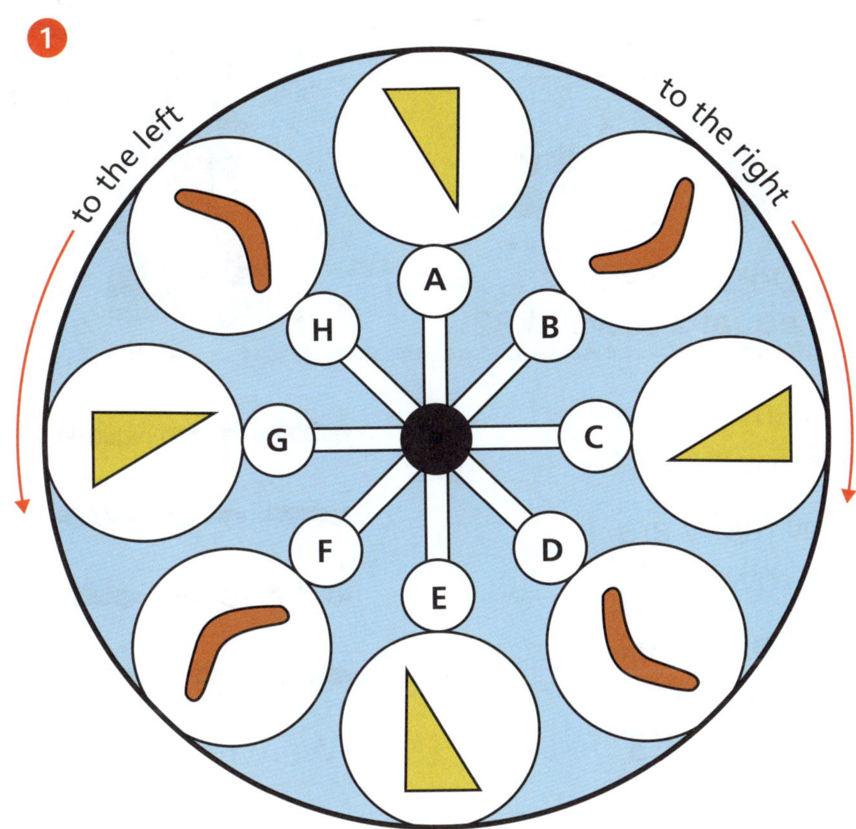

a When **A** turns a quarter turn to the right, it becomes shape ☐.

b When **A** turns a quarter turn to the left, it becomes shape ☐.

c When **A** turns a half turn to the right, it becomes shape ☐.

d When **B** turns a half turn to the right, it becomes shape ☐.

e When **B** turns a quarter turn to the right, it becomes shape ☐.

Using pattern blocks

ACTIVITY

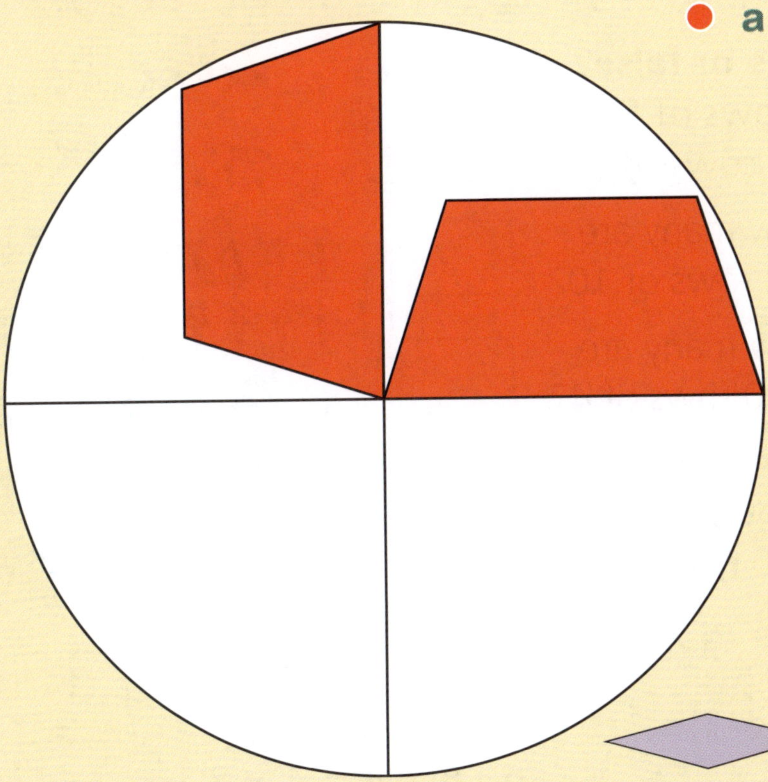

- **a** Use a pattern block to finish this pattern. Turn it around the centre and trace the block.
  - **b** Make other turning patterns using the pattern blocks.
- Use pattern blocks to make patterns by flipping and sliding.

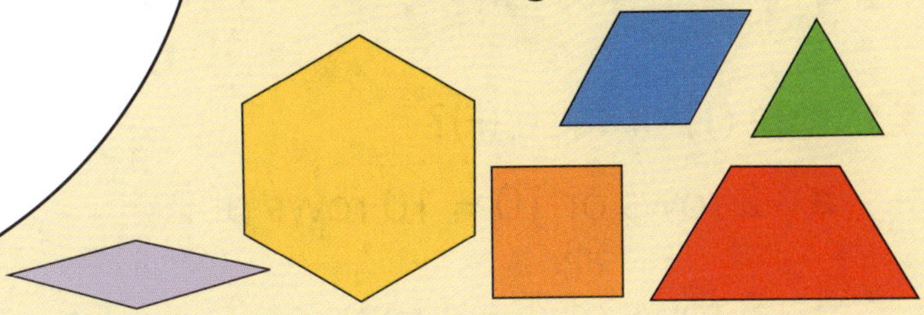

26A Using rows

4 rows of 3 = 12
We write 4 x 3 = 12.

3 rows of 4 = 12
We write 3 x 4 = 12.

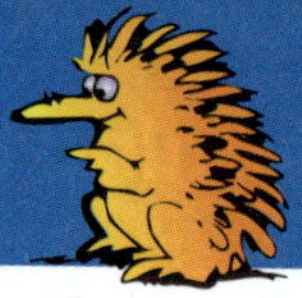

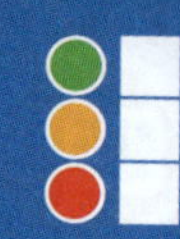

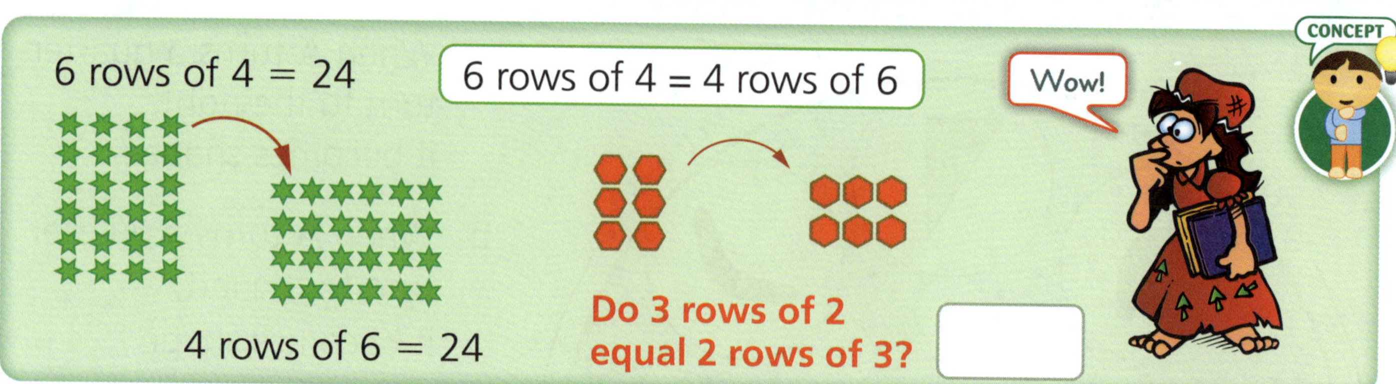

1 How many are in 3 rows of 4?

How many are in 4 rows of 3?

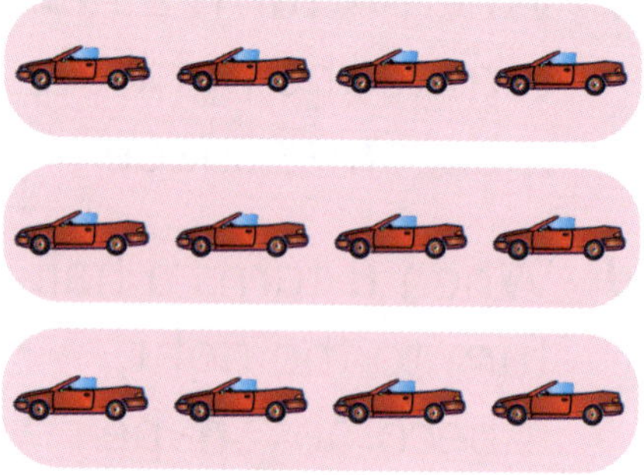

True or false?
3 rows of 4 = 4 rows of 3

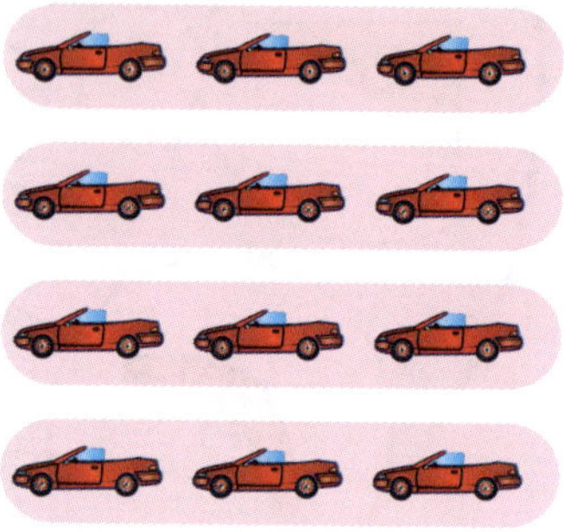

We write 3 × 4

We write 4 × 3

2 How many are in 3 rows of 5?

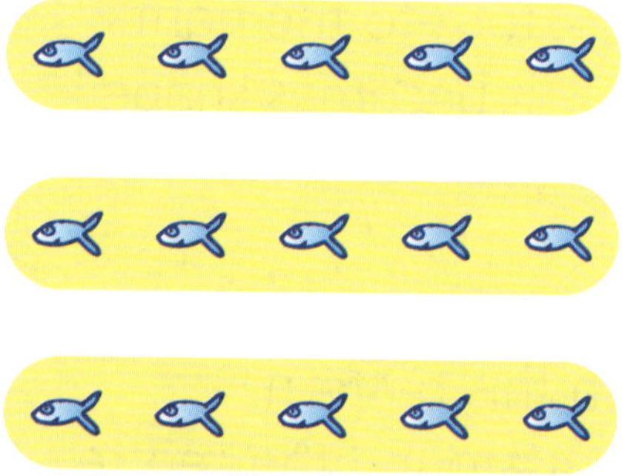

How many are in 5 rows of 3?

True or false?
3 rows of 5 = 5 rows of 3

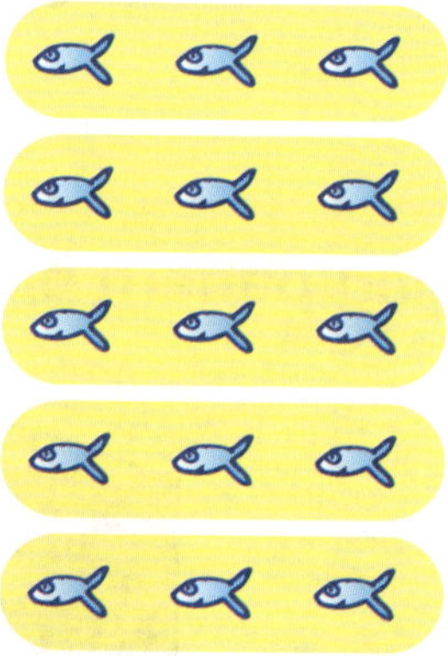

We write 3 × 5

3 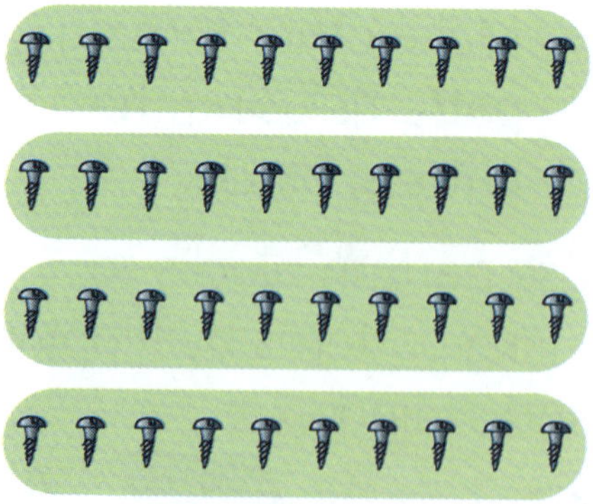 How many are in 4 rows of 10?

How many are in 10 rows of 4?

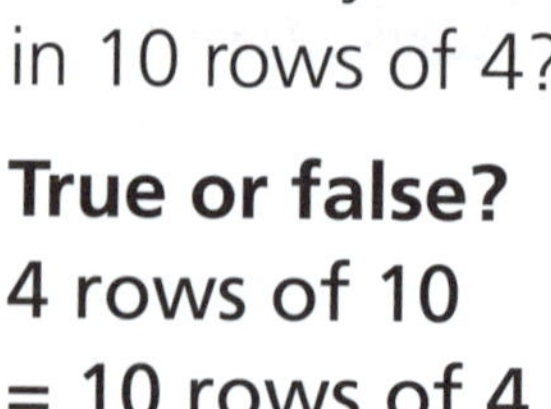

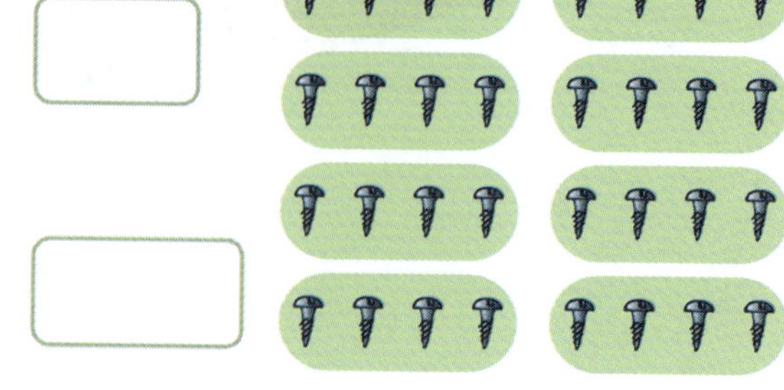

True or false?
4 rows of 10 = 10 rows of 4

4 True (T) or false (F)?

a 2 rows of 10 = 10 rows of 2 

b 2 × 10 = 10 × 2

c 8 rows of 4 = 4 rows of 8 

d 8 × 4 = 4 × 8 

Using groups

2 groups of 6
2 x 6

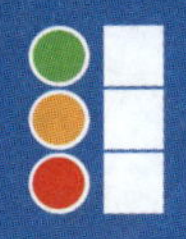

CONCEPT

We can use counters or blocks to model problems.

I had 20 stars. I put 5 on each card.
How many cards did I use?

20			
5			

Answer: I used 4 cards.

Get well soon.

Circle rows of 5 stars.

1 Draw circles to show these groups and rows.

a groups of 3 balls

How many balls? ☐

How many groups? ☐

How many balls in each group? ☐

b rows of 6 flowers

How many flowers? ☐

How many rows? ☐

How many flowers in each row? ☐

c groups of 5 fish

How many fish? ☐

How many groups? ☐

How many fish in each group? ☐

d groups of 2 coins

How many coins? ☐

How many groups? ☐

How many coins in each group? ☐

e groups of 4 stars

How many stars? ☐

How many groups? ☐

How many stars in each group? ☐

f rows of 5 coins

How many coins? ☐

How many rows? ☐

How many coins in each row? ☐

INVESTIGATION

Use 24 counters. How many students can be given:

a 2 counters? ☐ **b** 3 counters? ☐ **c** 4 counters? ☐

d 6 counters? ☐ **e** 8 counters? ☐ **f** 1 counter? ☐

26C Adding columns

three columns of 2
We say 3 x 2.

CONCEPT

4 columns of 6
(4 × 6)
number of stars ☐ 

4 rows of 6
(4 × 6) 
number of stars ☐

4 groups of 6
(4 × 6) 
number of stars ☐

True (T) or false (F)? 4 columns of 6 = 4 rows of 6 = 4 groups of 6 ☐

1 a 3 columns of 4 (3 × 4) ☐

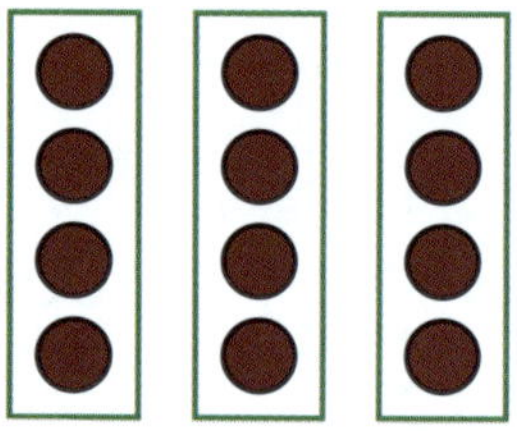

b 4 columns of 4 (4 × 4) ☐

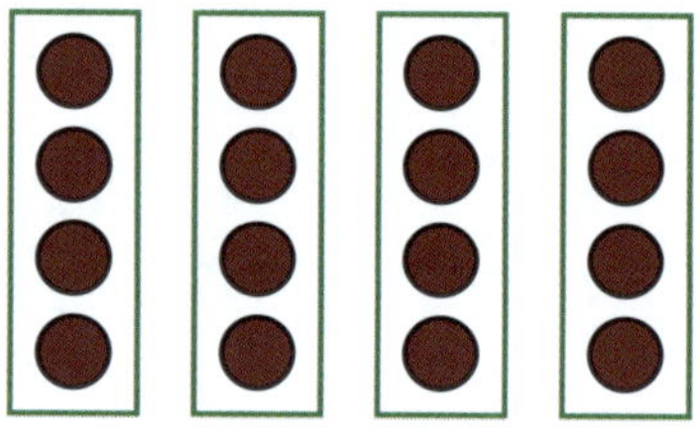

c 5 columns of 4 (5 × 4) ☐

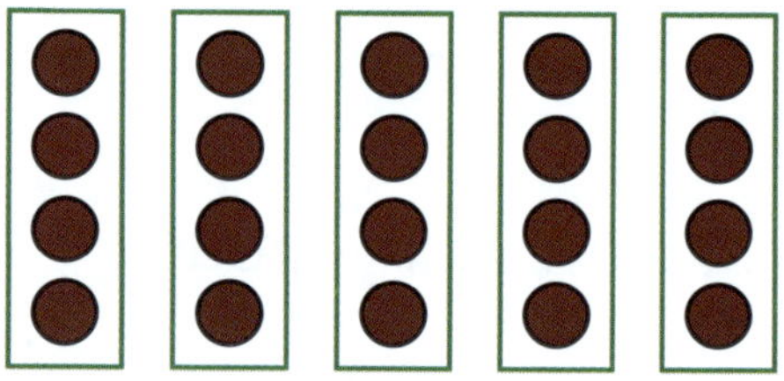

d 4 columns of 5 (4 × 5) ☐

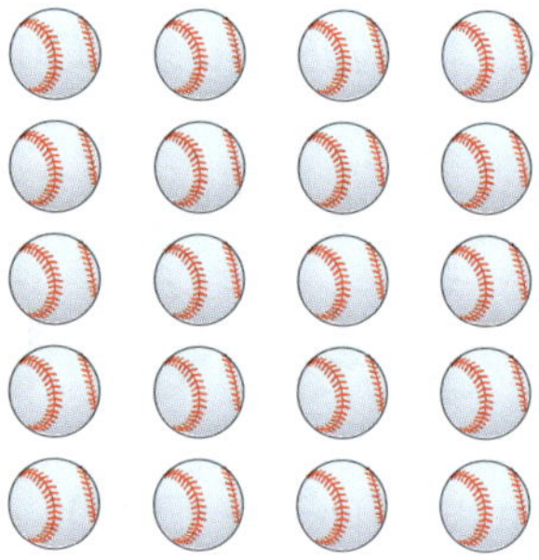

e 5 columns of 5 (5 × 5) ☐

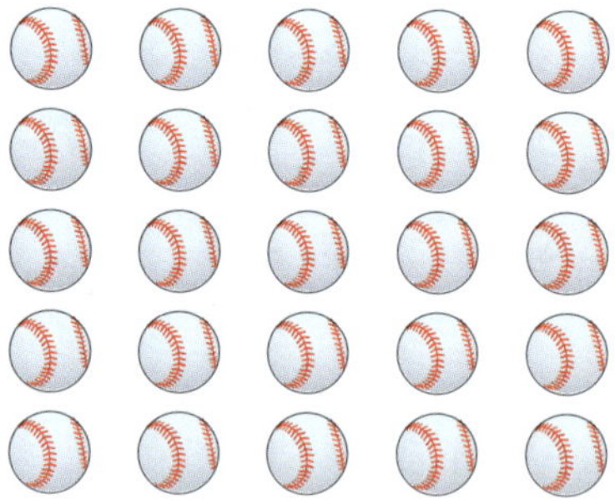

f 6 columns of 5 (6 × 5) ☐

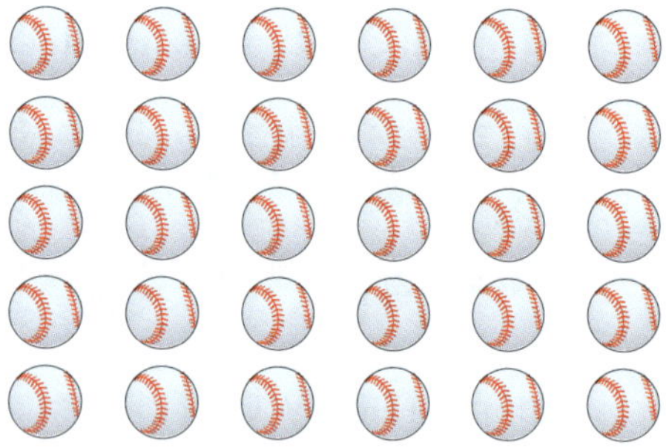

2

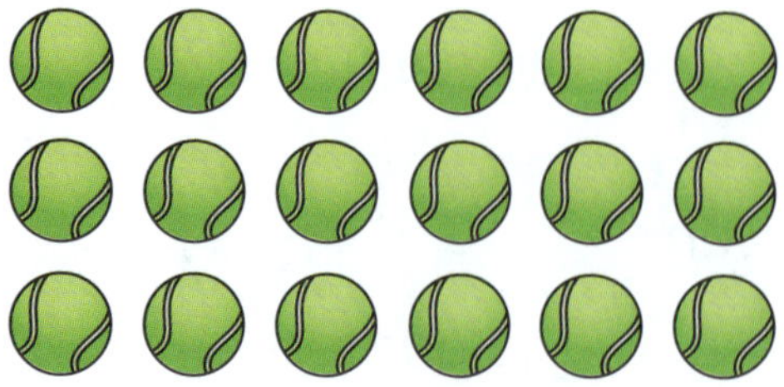

Skip count: 3, 6, 9, 12, 15 ...

a How many columns of 3? ☐

b How many rows of 6? ☐

c How many groups of 2? ☐

d How many balls altogether? ☐

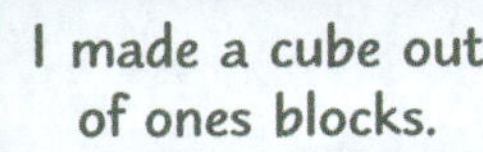

This is a cube.

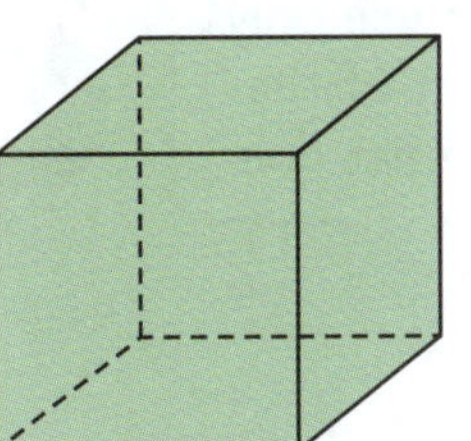

1 Circle the cubes.

die

box of soap

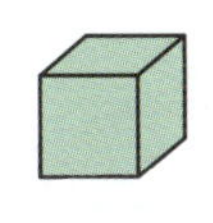

ones block

2 **vertex (or corner)**

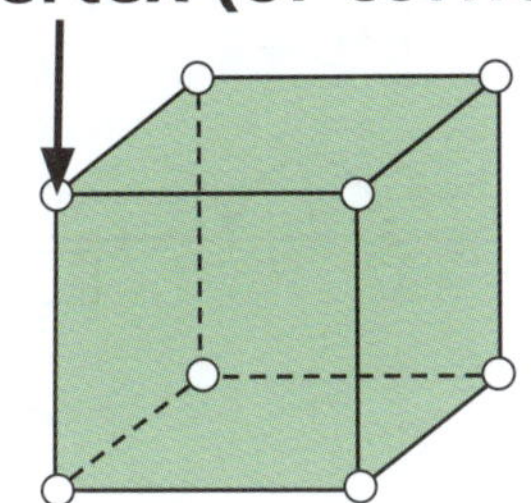

Corners are called vertices.
A vertex is one corner.

Colour each vertex on this cube.

How many vertices does a cube have? ☐

3 **face**

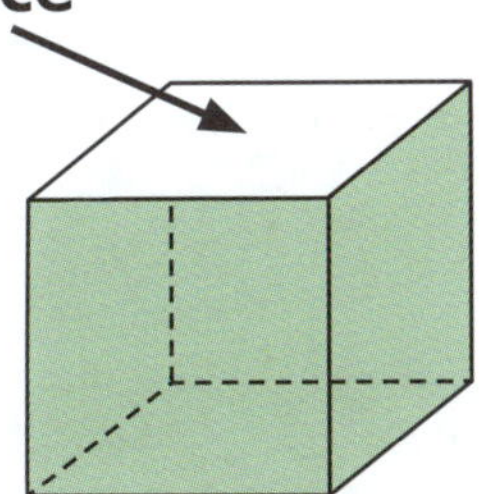

A face of an object is a flat surface that has straight sides.

How many faces does a cube have ☐

4 **edge**

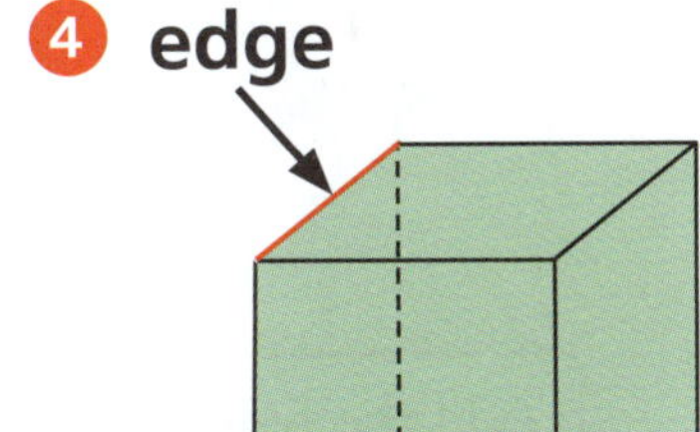

An edge is where two surfaces meet. Two faces will meet in a straight edge.

Trace over each edge of this cube.

How many edges does a cube have? ☐

5 **a** What shape is the face of a cube? ☐

b How many of these shapes are found on a cube? ☐

Use plasticine or playdough to make a model of a cube.

27A Jump strategy (addition)

38 + 6
= 38 + 2 + 4

CONCEPT

38 + 6 Start at 38 and jump to 40.

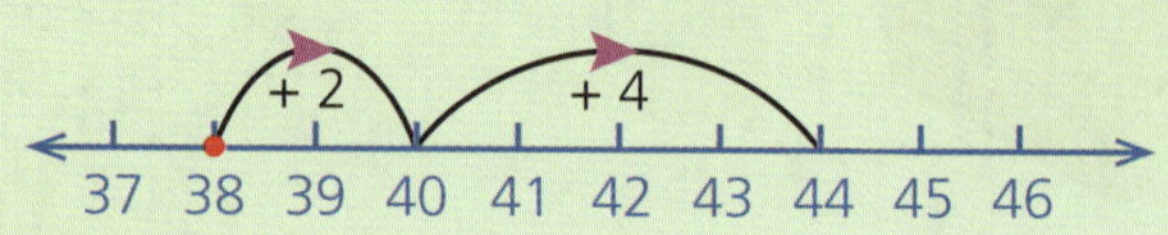

$$\begin{array}{r} 38 \\ +\ 6 \\ \hline 44 \end{array}$$

1 Use the jump strategy to find the answers.

a 38 + 4 = ☐

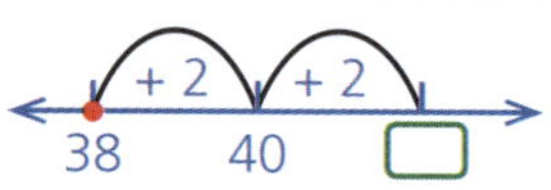

b 58 + 5 = ☐

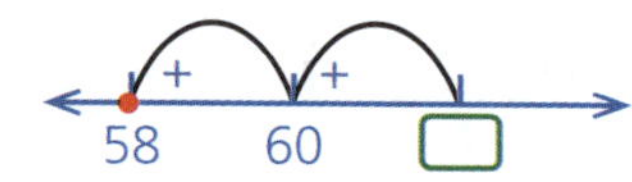

c 47 + 6 = ☐

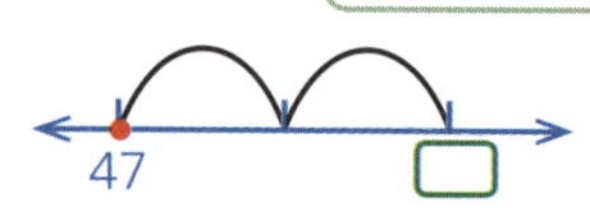

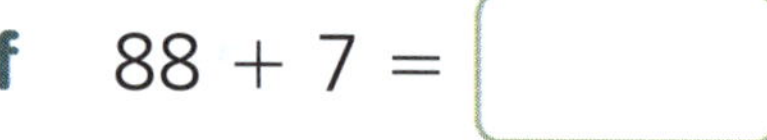

d 37 + 8 = ☐

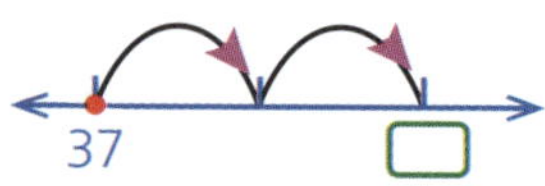

e 57 + 9 = ☐

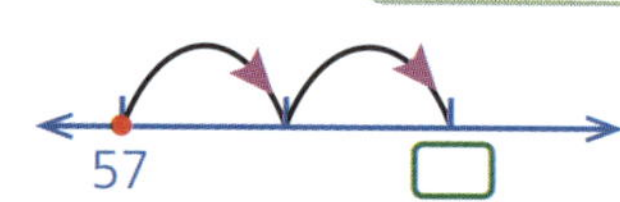

f 88 + 7 = ☐

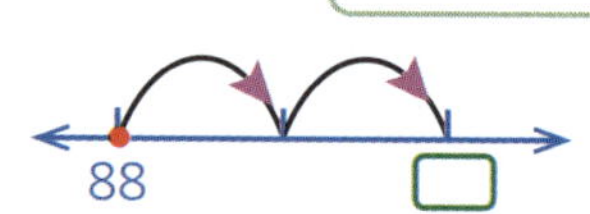

CONCEPT

37 + 24 Start at 37 and count on.

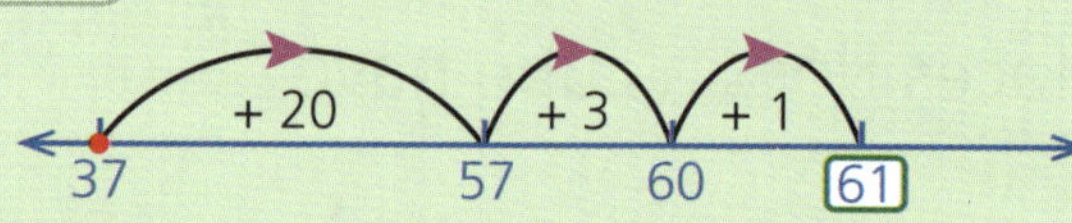

- We don't need marks on the line.
- We jump the tens, then the ones.

or

37 + 24
= 37 + 20 + 4
= 57 + 3 + 1
= 61

$$\begin{array}{r} 37 \\ +\ 24 \\ \hline 61 \end{array}$$

2 Use the jump strategy to find the answers.

a 46 + 21 = ☐

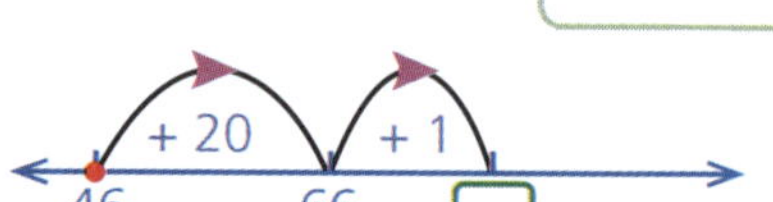

b 39 + 23 = ☐

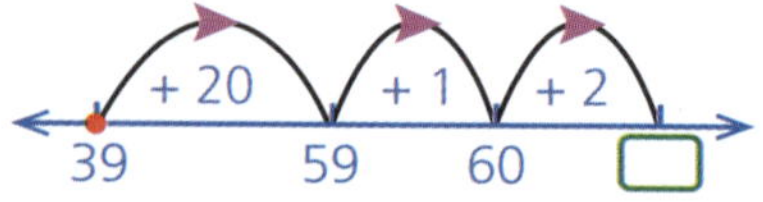

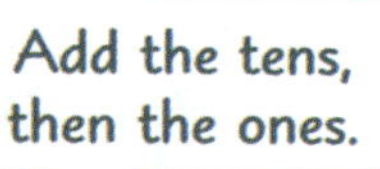

c 58 + 14 = ☐

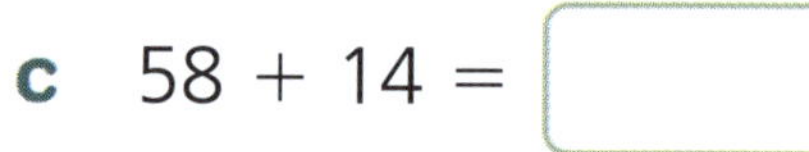

d 67 + 13 = ☐

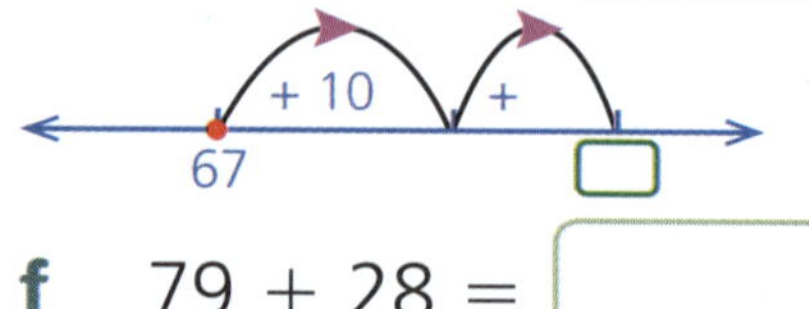

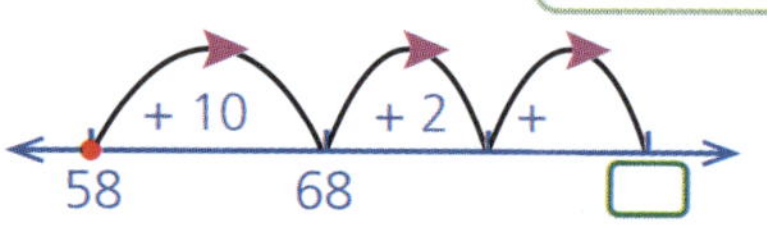

e 55 + 27 = ☐

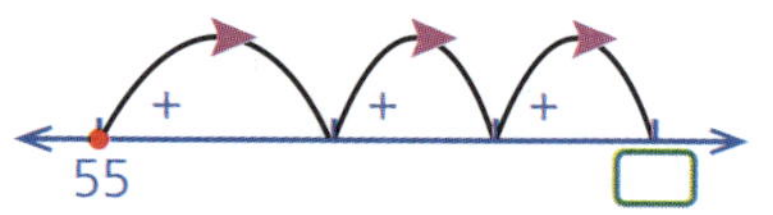

f 79 + 28 = ☐

+ + +
79

 ISBN 9780655709039

27B Jump strategy (subtraction)

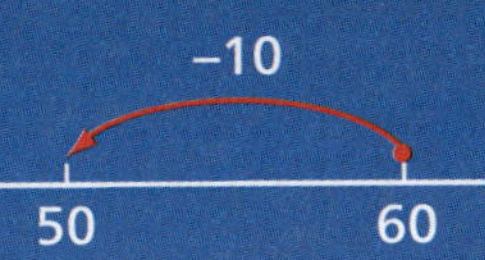

63 − 24 Start at 63 and jump back.

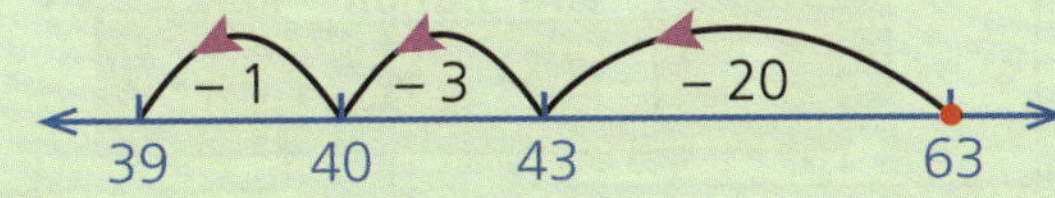

- We don't need marks on the line.
- We jump the tens, then the ones.

63 − 24 = 39

Use the jump strategy to find the answers.

1 a 52 − 5 = ☐ b 31 − 6 = ☐ c 93 − 4 = ☐

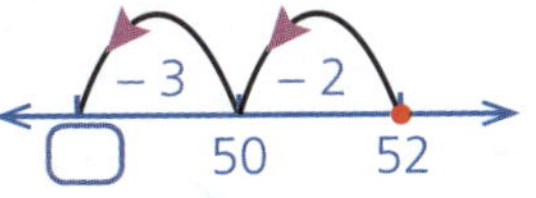

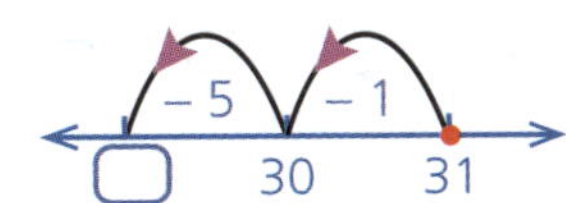

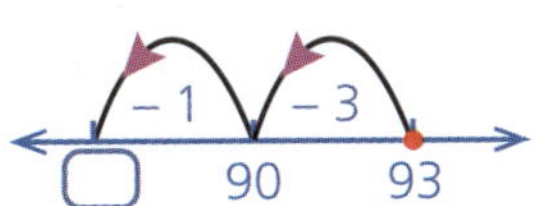

d 37 − 8 = ☐ e 57 − 9 = ☐ f 88 − 7 = ☐

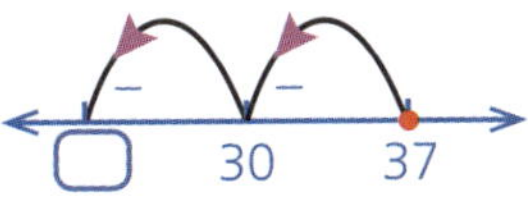

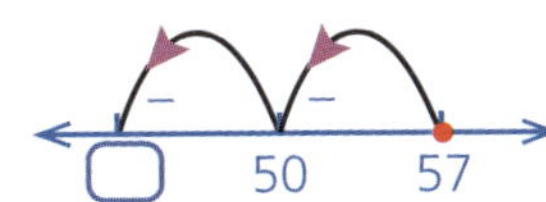

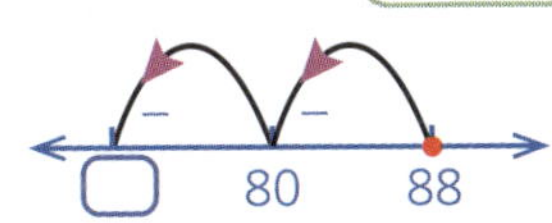

2 a 43 − 14 = ☐ b 64 − 25 = ☐

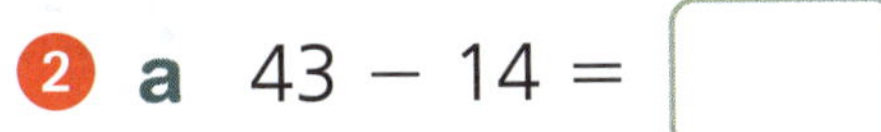
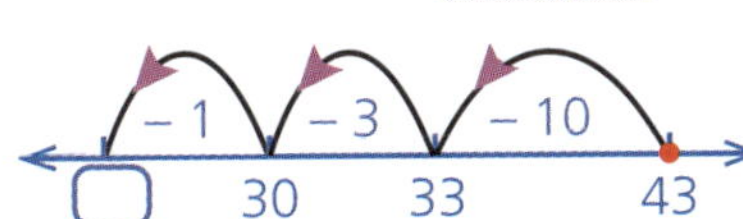

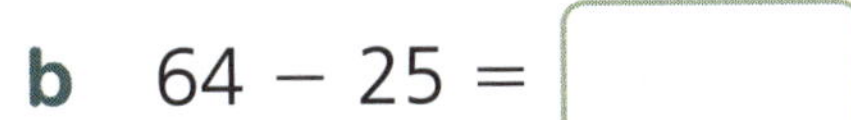
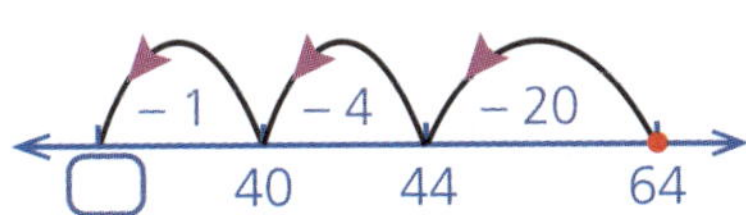

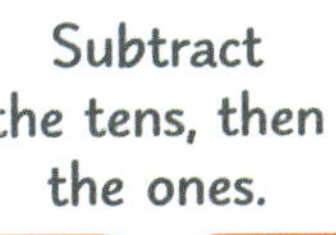

c 71 − 33 = ☐ d 82 − 57 = ☐

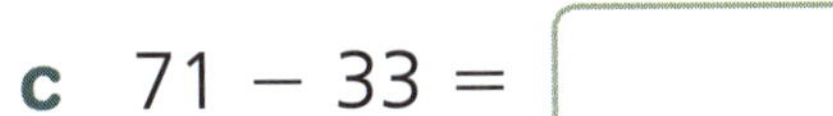
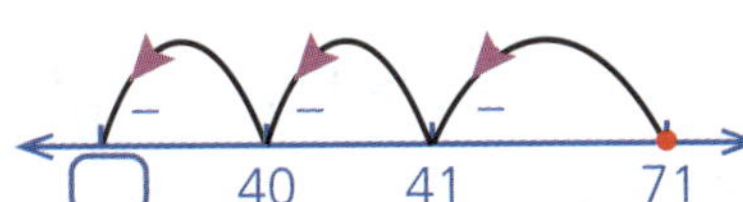

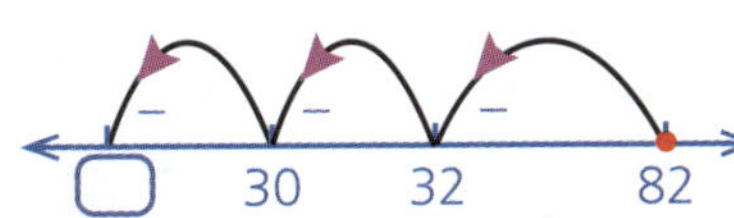

e 92 − 46 = ☐ f 55 − 38 = ☐

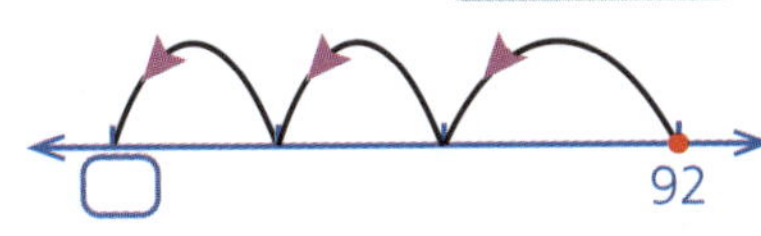

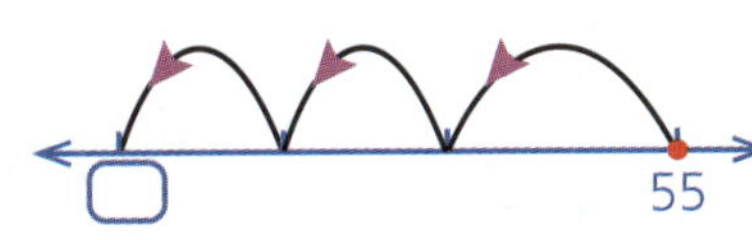

3 Can you do these in your head? If not, use a number line.

a 72 − 37 = ☐ b 96 − 29 = ☐ c 64 − 46 = ☐

d 85 − 27 = ☐ e 31 − 17 = ☐ f 42 − 23 = ☐

27C Giving directions

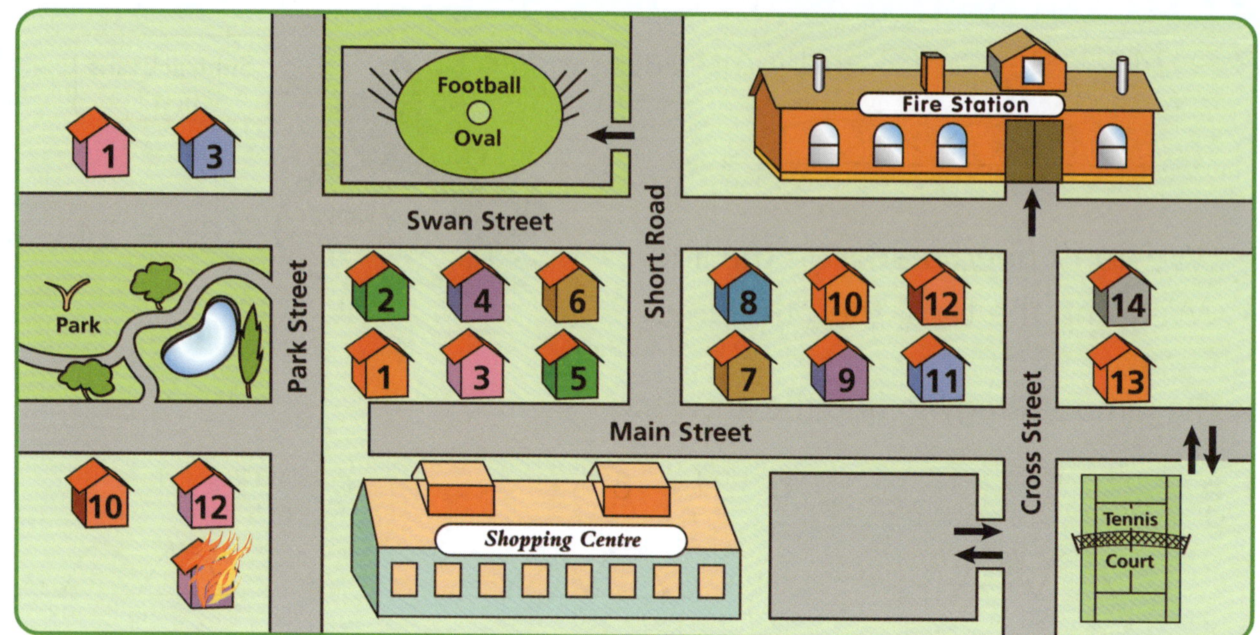

a Trace the shortest route from the fire station to the burning house. Write instructions.

____

____

b Write instructions for getting from 11 Main Street to the football oval.

____

____

FUN SPOT

Billy Ahmed Cooper

salmon shark tuna

Follow each line until you come to a fish. What did they catch?

Billy	
Ahmed	
Cooper	

27D Using tally marks

|||||| = 𝍸|

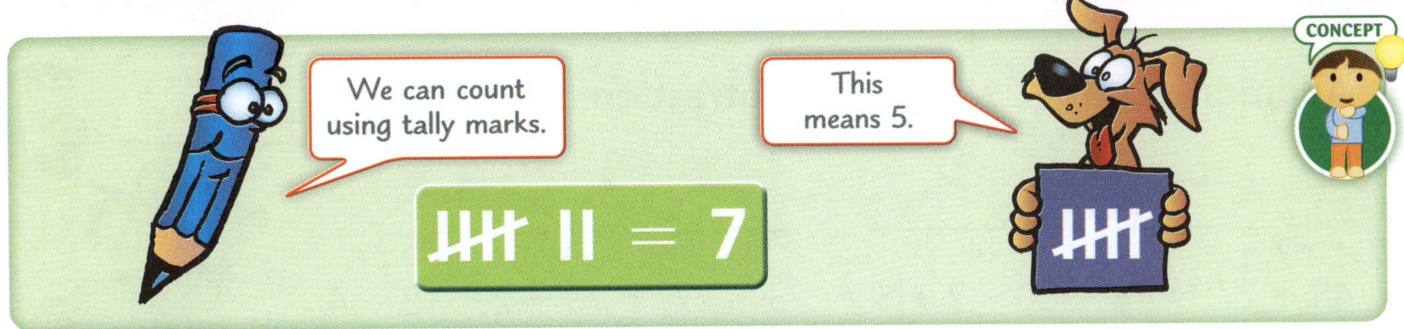

1 Write the number shown by each tally.

a ||| ☐ **b** 𝍸 | ☐ **c** 𝍸 𝍸 ☐

d 𝍸 |||| ☐ **e** 𝍸 𝍸 | ☐ **f** 𝍸 | ☐

2 Cross off one ball at a time and put a tally mark in the table.

Christmas balls	Number
red	
blue	
green	
orange	

O G G R G B G
G B G G B G R
B G R R G G B
R G O G B R G
G B G R G G R
G R R G G O G

How many of the balls are:

a red? ☐ **b** blue? ☐ **c** green? ☐

d orange? ☐ **e** How many balls altogether? ☐

Make a tally

Complete this tally for the people in your class. You can decide who has light hair and who has dark hair.

Hair colour	Number
light	
dark	

28A Jump strategy

47 + 10 = 57 67 − 10 = 57
47 + 20 = 67 67 − 20 = 47
47 + 30 = 77 67 − 30 = 37

CONCEPT

Addition 48 + 27

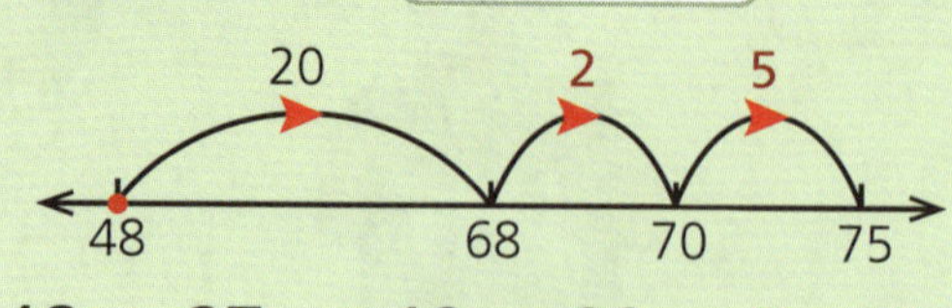

$48 + 27 = 48 + 20 + 2 + 5$
$= 68 + 2 + 5$
$= 70 + 5 = 75$

Subtraction  73 − 27

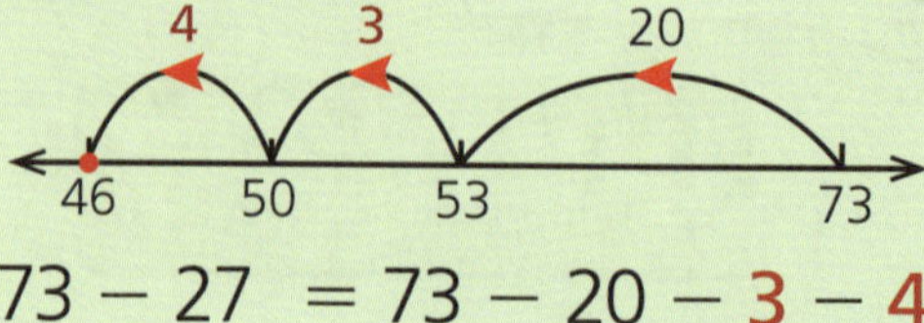

$73 - 27 = 73 - 20 - 3 - 4$
$= 53 - 3 - 4$
$= 50 - 4 = 46$

1 **a** 42 + 19 ☐ (42)

b 35 + 48 ☐ (35)

c 39 + 45 ☐

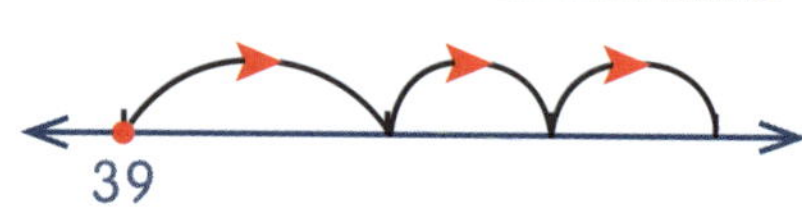

d 27 + 28 ☐

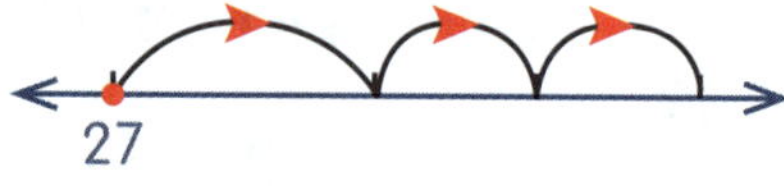

e 73 + 22 ☐

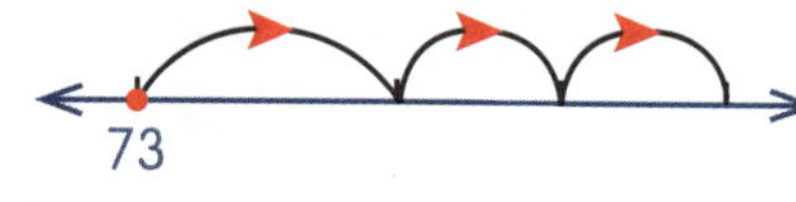

f 85 + 18 ☐

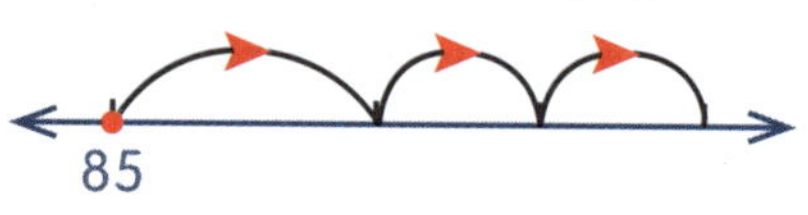

g 47 + 35 ☐ (47)

h 86 + 29 ☐

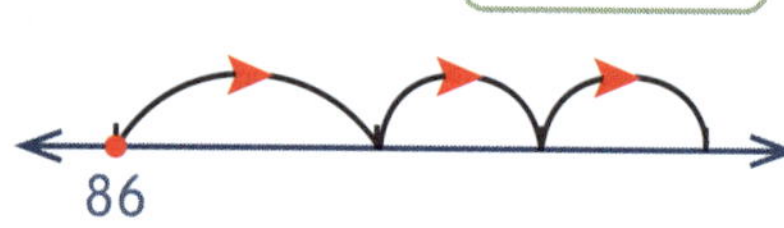

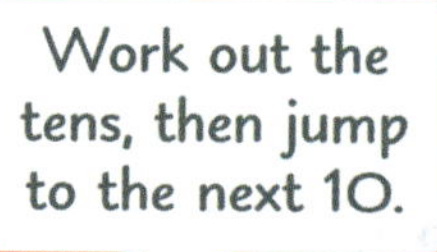

2 **a** 42 − 14 ☐

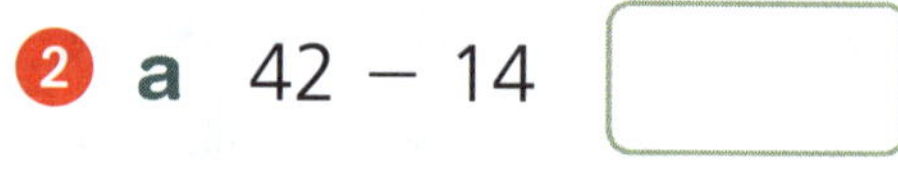

b 95 − 26 ☐

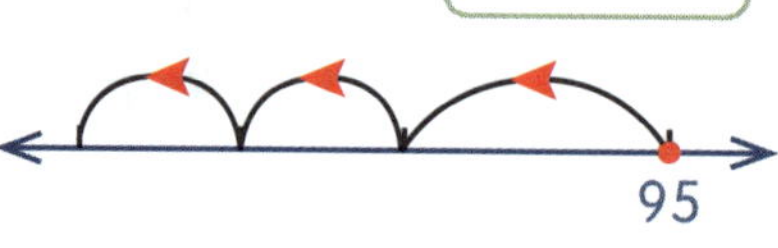

c 52 − 23 ☐

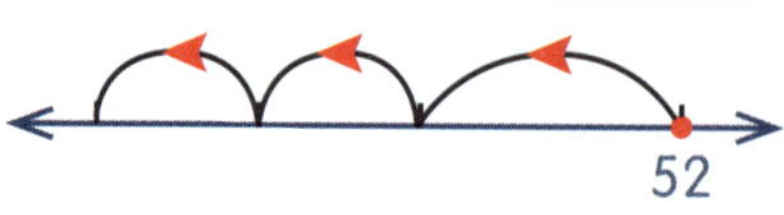

d 73 − 22 ☐

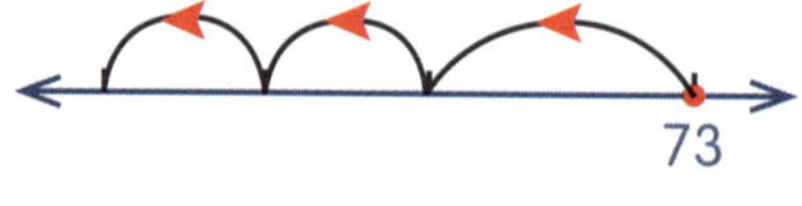

e 85 − 18 ☐

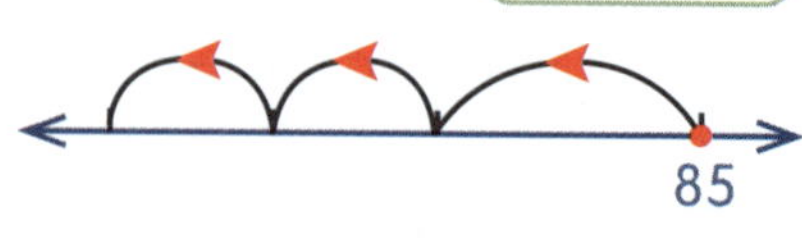

f 66 − 27 ☐

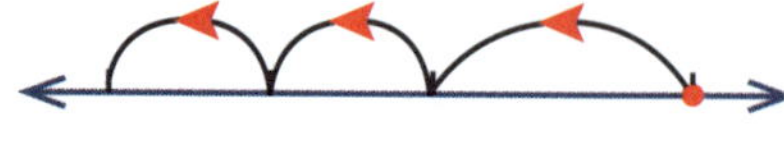

g 88 − 65 ☐

h 94 − 59 ☐

3 **a** 38 + 25 ☐ **b** 86 − 29 ☐ **c** 19 + 42 ☐

 • *AUSTRALIAN SIGNPOST MATHS NSW 2* • ISBN 9780655709039

28B Fractions of a group

CONCEPT

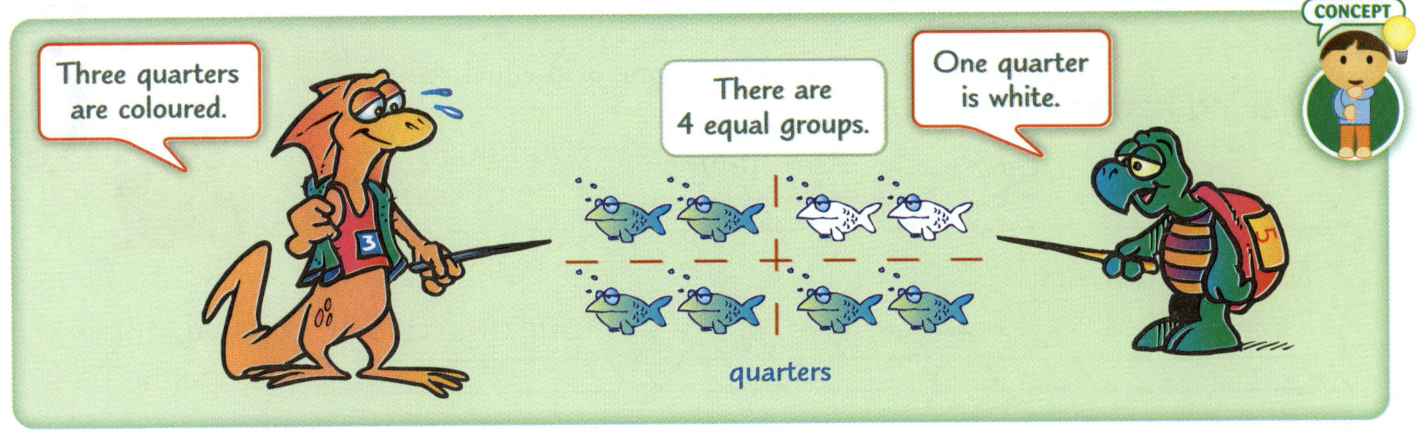

1 Circle the groups that are divided into quarters and colour one quarter.

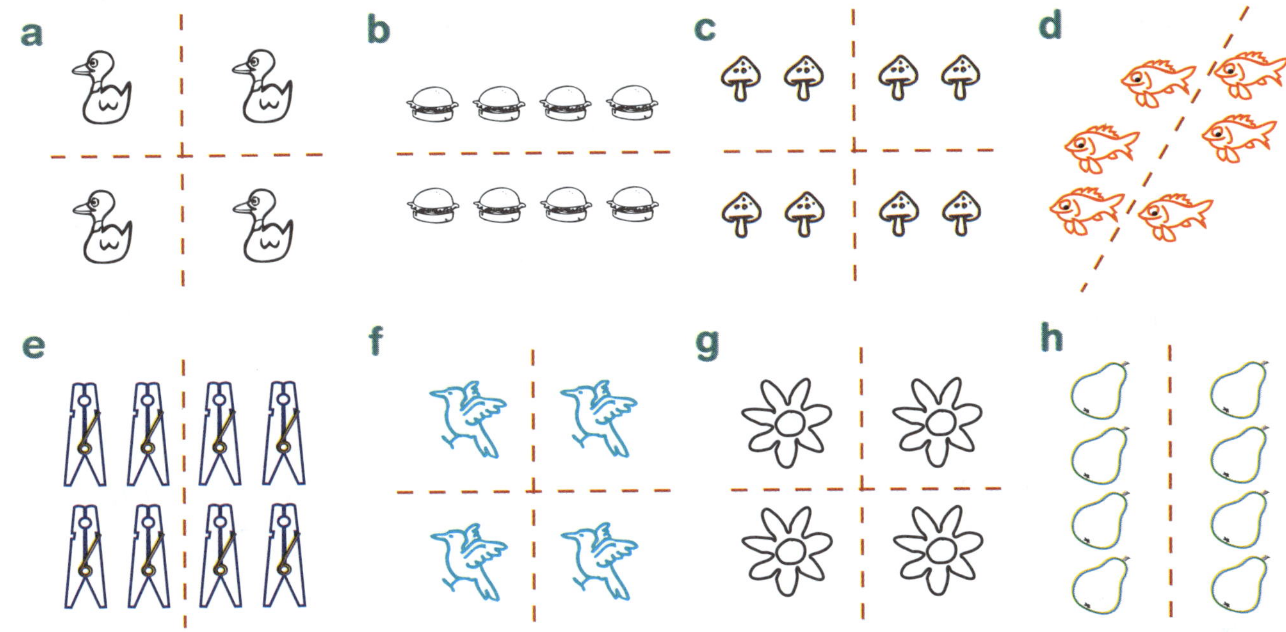

INVESTIGATION

Here we have displayed 12 counters, as quarters of a whole.

Use 8 counters to make a model like this and explain how you made your model.

28C Eighths of a length

One eighth of the beads is coloured blue.

CONCEPT

One eighth is one of 8 equal parts.

1 Making halves

two halves

We folded the strip of paper into two halves.

2 Making quarters

two halves

four quarters

We have folded each half in half.

3 Making eighths

two halves

There are eight equal parts.

eight eighths

We have halved this strip of paper 3 times.

Put 8 ones blocks together.

Each block shows one eighth of the length.

1 Use this 8 cm strip to colour:

a one eighth of the length

b two eighths of the length

c three eighths of the length

d four eighths of the length

e five eighths of the length

f six eighths of the length

g seven eighths of the length

h eight eighths of the length

2 **a** One half is the same as ☐ eighths.

b One whole is the same as ☐ eighths.

28D Making graphs

||||| = 𝍸 or 𝍸

Without counting, estimate which group is largest.

Comparing groups is fun.

1 Complete this table to record the animals in the picture. (Cross out each animal as you put its tally mark in the table.)

Animals seen	Tally	Number
Birds		
Kangaroos		
Wombats		

2 On the graph, draw one picture for each animal.

a Which group is the smallest?

b Which group is the largest?

c How many more birds are there than wombats?

d What is the difference between the number of kangaroos and the number of birds?

Animals seen		
Birds	Kangaroos	Wombats

INVESTIGATION

Draw a table showing the people in your classroom. Circle the category you predict will be largest.

Boys	
Girls	
Adults	

 • *AUSTRALIAN SIGNPOST MATHS NSW 2* • ISBN 9780655709039

29A Doubling and halving

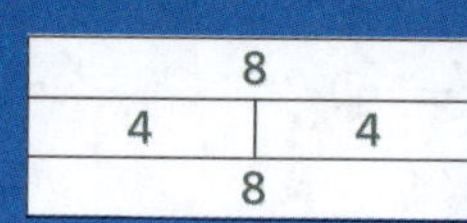

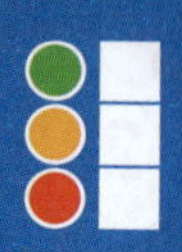

1 Half of the counters are covered with paper.
Draw the missing counters to double each group.

a Double 2 = ☐
Half of 4 = ☐

b 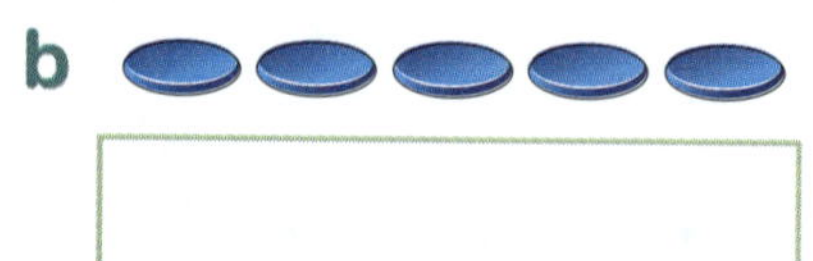
Double 5 = ☐
Half of 10 = ☐

c Double 3 = ☐
Half of 6 = ☐

d Double 4 = ☐
Half of 8 = ☐

2 a Half of Levi's blocks are shown. Draw the other half below the ones shown.

When the half is doubled there are ☐.
Half of the whole group is ☐.

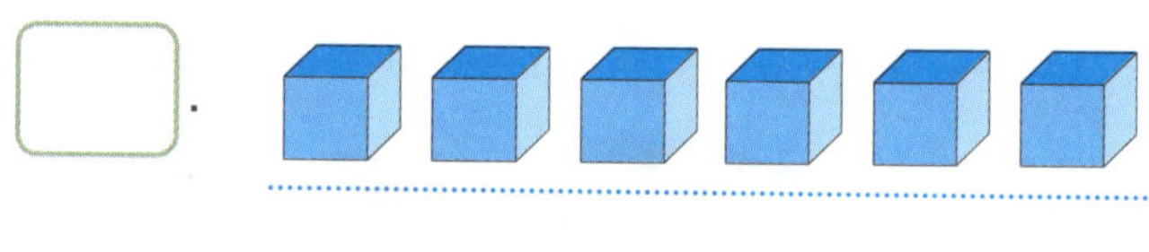

b Half of Sally's ducks are shown. Draw the other half below the ones shown.

When the half is doubled there are ☐.
Half of the whole group is ☐.

c Half of Ali's coins are shown. Draw the other half below the ones shown.

When the half is doubled there are ☐.
Half of the whole group is ☐.

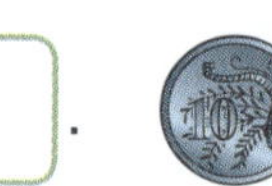

3 Halve 10. ☐ Add 3 to your answer. ☐ Double this. ☐
Halve your last answer. ☐ Halve your last answer. ☐ Double this. ☐

Choose an even number of counters. Cover half of your group with paper. Ask your partner to find the total number of counters.

29B Doubling and halving

Halve 10 then double.

10	
5	5
10	

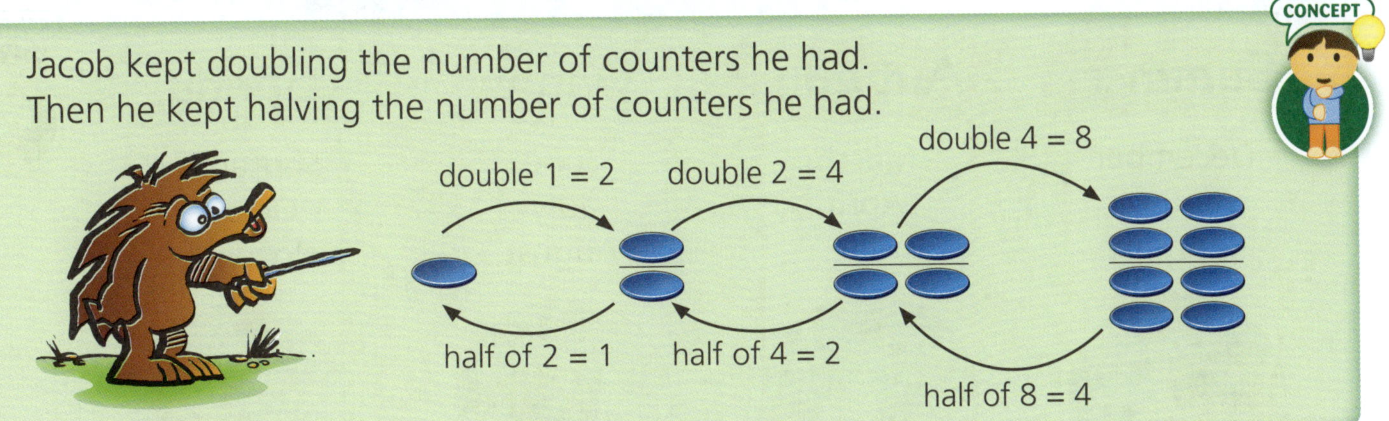

1 Draw pictures to double the number of shapes.
Then cross out half of the answer. Write what you found.

a

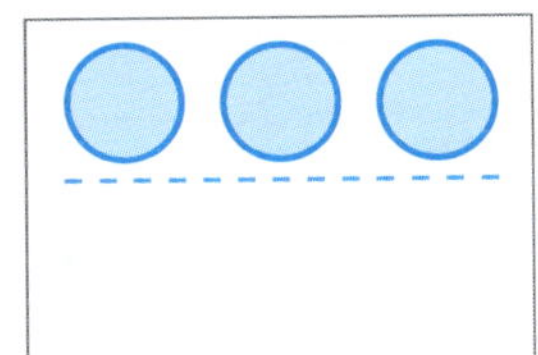

Double 3 = ☐

Half of  = ☐

b

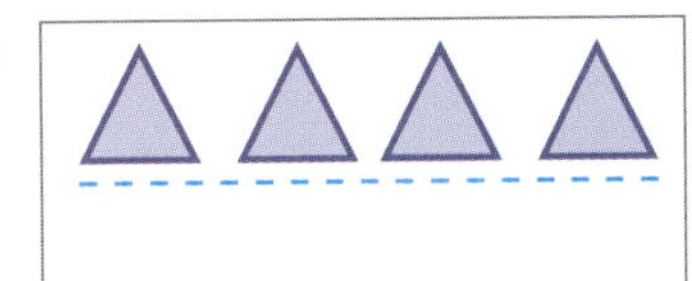

Double 4 = ☐

Half of ☐ = ☐

If you double a number and halve the result, the number …

☐

☐

2 Half my money is shown. Draw the rest, and write the total amount of my money.

a

Total = ☐

b

Total = ☐

3 I doubled my money then spent half of the total. Draw what I did.

a

Double $1 = ☐

Half of the last answer = ☐.

b

Double $2 = ☐

Half of the last answer = ☐.

29C Duration of time

CONCEPT

Summer	Autumn	Winter	Spring
December January February	March April May	June July August	September October November

1. How many months is it from the beginning of summer to the:

   a end of autumn? ☐ **b** beginning of autumn? ☐

   c end of spring? ☐ **d** beginning of spring? ☐

July

S	M	T	W	T	F	S
				1	2	3
4	5	6	7	8	9	10
11	12	13	14	15	16	17
18	19	20	21	22	23	24
25	26	27	28	29	30	31

August

S	M	T	W	T	F	S
1	2	3	4	5	6	7
8	9	10	11	12	13	14
15	16	17	18	19	20	21
22	23	24	25	26	27	28
29	30	31				

September

S	M	T	W	T	F	S
			1	2	3	4
5	6	7	8	9	10	11
12	13	14	15	16	17	18
19	20	21	22	23	24	25
26	27	28	29	30		

October

S	M	T	W	T	F	S
					1	2
3	4	5	6	7	8	9
10	11	12	13	14	15	16
17	18	19	20	21	22	23
24	25	26	27	28	29	30
31						

November

S	M	T	W	T	F	S
	1	2	3	4	5	6
7	8	9	10	11	12	13
14	15	16	17	18	19	20
21	22	23	24	25	26	27
28	29	30				

December

S	M	T	W	T	F	S
			1	2	3	4
5	6	7	8	9	10	11
12	13	14	15	16	17	18
19	20	21	22	23	24	25
26	27	28	29	30	31	

2. How many months and days is it from 5th July to:

   a 7th September? ☐ months ☐ days

   b 25th August? ☐ months ☐ days

   c 10th December? ☐ months ☐ days

   d 21st November? ☐ months ☐ days

3. How many weeks and days is it from 5th July to:

   a 7th September? ☐ weeks ☐ days

   b 25th August? ☐ weeks ☐ days

   c 10th December? ☐ weeks ☐ days

   d 21st November? ☐ weeks ☐ days

Some Indigenous people around Australia use six seasons. The six major seasons of the Indigenous people in Arnhem Land, Northern Territory are called:

Dhuludur, Barramirri, Mayaltha, Midawarr, Dharratharramirri and Rarrandharr.

 • *AUSTRALIAN SIGNPOST MATHS NSW 2* • ISBN 9780655709039

Gathering data

1 Rachel recorded the girls (**G**) and the boys (**B**) arriving at her party. This is the list she made. Use tally marks to fill in the table.

G G G B B B B G B G G G B B B
G B B B G G B G B G B B G G B

	Tally	Number
Girls		
Boys		

Make a graph to show how many boys and girls came to the party by colouring one picture for each person.

Friends at the party

2 a Write true (T) or false (F) next to each number sentence.

1 + 1 = 8	F	7 + 1 = 8		2 + 2 = 5		3 + 3 = 6	
1 + 5 = 6		2 + 5 = 9		2 + 3 = 5		5 + 1 = 6	
2 + 1 = 3		4 + 4 = 2		8 + 2 = 9		7 + 5 = 8	
4 + 5 = 6		8 + 0 = 8		0 + 9 = 9		2 + 5 = 7	

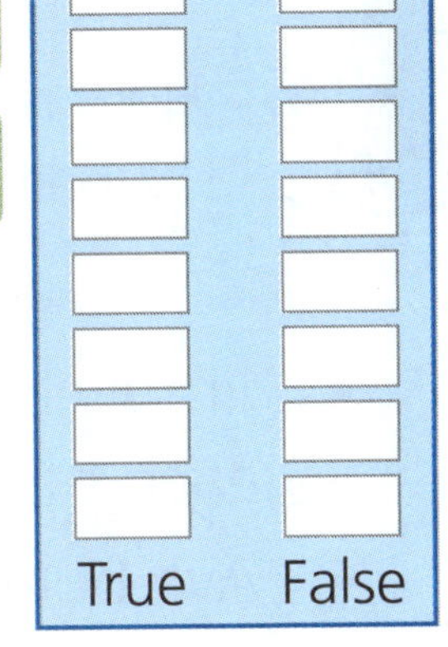

b Complete the table and draw a graph by colouring boxes.

	Tally	Number
True		
False		

c How many questions were there altogether?

 • *AUSTRALIAN SIGNPOST MATHS NSW 2* • ISBN 9780655709039

CONCEPT

Jan made 8 towers with 4 blocks in each tower.
How many blocks did she use?

4, 8, 12, 16, 20, 24, 28, 32 …

☐ columns of 4

Jan used 32 blocks.

8 columns of 4 = 4 + 4 + 4 + 4 + 4 + 4 + 4 + 4 = ☐

1 There are 4 blocks in each tower. (You can use the skip counting pattern above.)

a	How many blocks are in 3 towers?	☐ columns of 4	There are ☐ blocks.
b	How many blocks are in 4 towers?	☐ columns of 4	There are ☐ blocks.
c	24 blocks were used to build towers. How many towers were built?	How many 4s in 24 blocks?	There are ☐ towers.
d	20 blocks were used to build towers. How many towers were built?	How many 4s in 20 blocks?	There are ☐ towers.

2 Each can holds 3 tennis balls. (You can use the skip counting pattern above.)

a	How many balls are in 7 cans?	☐ groups of 3	There are ☐ balls.
b	How many balls are in 6 cans?	☐ groups of 3	There are ☐ balls.
c	How many cans can we fill with 12 balls?	How many 3s in 12 balls?	There are ☐ cans.

INVESTIGATION

- Jimmy bought 8 cans of 3 tennis balls. He gave 2 balls to each of his friends. How many friends were given balls ? ☐
- Ava bought 5 boxes of 4 cakes. She gave 3 cakes to her sister and 8 cakes to her brother. How many cakes did she have left? ☐

Problem solving

CONCEPT

John made 6 towers with 5 blocks in each tower.
How many blocks did he use?

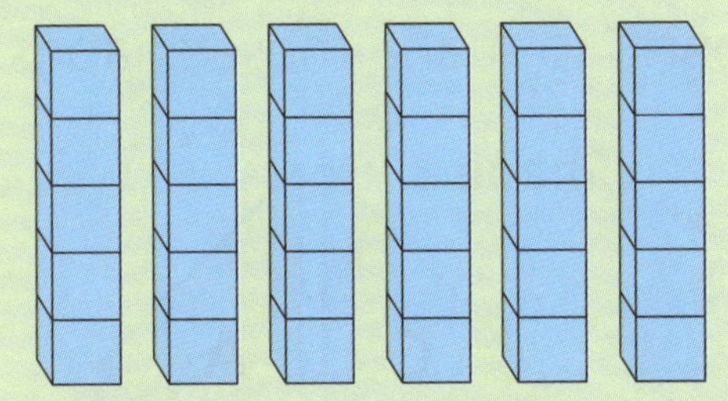

5, 10, 15, 20, 25, 30

John used 30 blocks.

6 columns of 5 = 5 + 5 + 5 + 5 + 5 + 5 = 30

Show working on your own paper.

1 **a** How many apples are in 3 bags if 3 apples are in each bag? — 3 bags of 3 apples. — There are ☐ apples.

b If there are 15 apples and 3 in each bag, how many bags? — How many 3s in 15 apples? — There are ☐ bags.

c You have 18 apples and put 3 in each bag. How many bags were used? — How many 3s in 18 apples? — There are ☐ bags.

2 Each tower is made of 5 blocks. (You can use the skip counting patterns above.)

a How many blocks are in 4 towers? — ☐ groups of 5 — There are ☐ blocks.

b How many blocks are in 6 towers? — ☐ groups of 5 — There are ☐ blocks.

c How many towers can be built using 20 blocks? — How many 5s in 20 blocks? — There are ☐ towers.

INVESTIGATION

Draw as many arrays as you can that use 24 tiles.
This array is 12 squares long and 2 squares wide.

Make sure that each array is a rectangle.

30C Combine and separate shapes

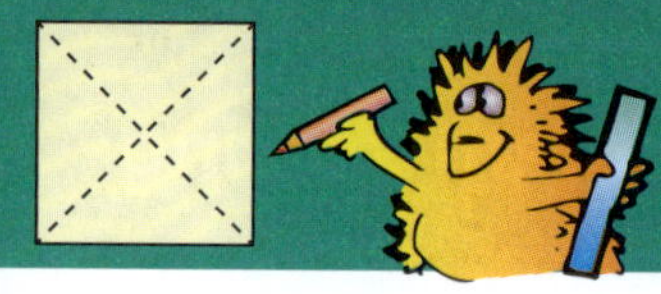

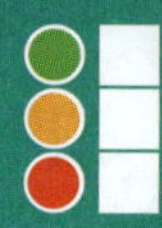

- Mary used a ruler to draw lines from corner to corner on her square paper.
- She then cut her square into 4 triangles and labelled them A, B, C and D.
- She coloured parts C and D, then used the 4 parts to make these shapes.

Shape 1

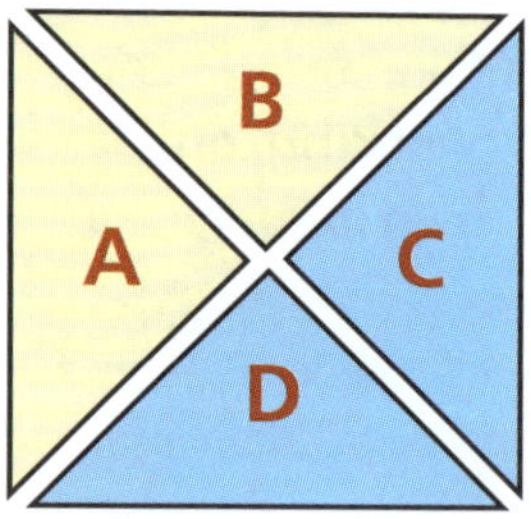

Shape 2

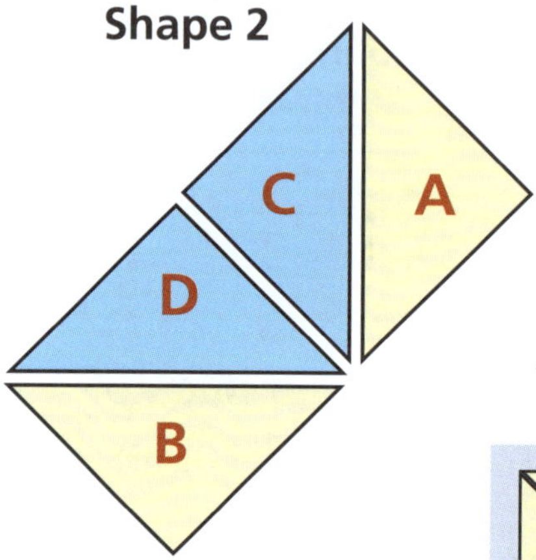

Shape 3

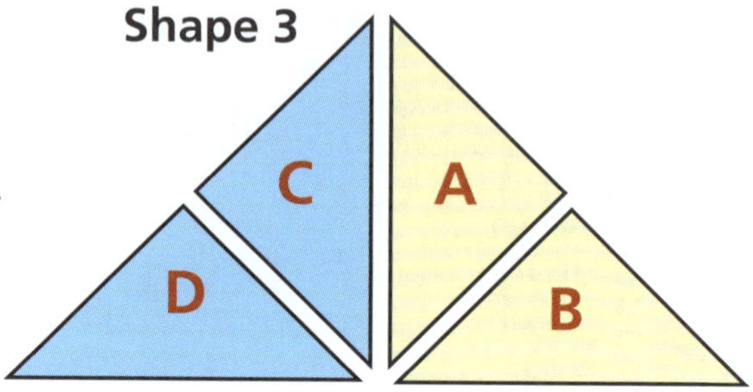

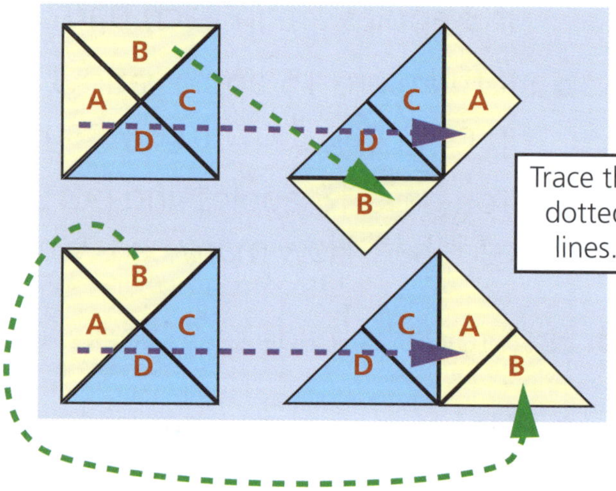

Trace the dotted lines.

1 What new shapes did Mary make?

How many ones blocks are needed to cover her square?

How many ones blocks are needed to cover her rectangle?

Do you think the square and the rectangle have the same area?

Would the triangle have the same area as the other two shapes?

Explain your answer.

ACTIVITY

- Use 1 cm grid paper.
- Use Mary's method to cut a square into 3 triangles. Use your 3 shapes to make a triangle and a rectangle.

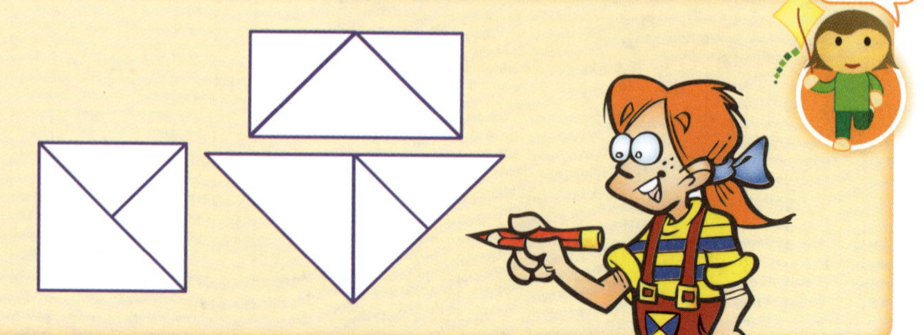

30D Following instructions

1. Little Red Riding Hood followed these instructions. Colour her path red.

Woodcutter

Grandma's house

Start here

Instructions

Move
2 left
3 up
1 left
1 down
3 left
3 up
6 right
2 up
4 left

2. The woodcutter moved 4 right, then 2 down, then 4 left. Draw a tree in that square.
3. Give directions for the woodcutter to move to the wolf. Colour his path blue.
4. Give directions for the wolf to move to the bird. Colour the path green.

Make up other trips using the grid.

 • *AUSTRALIAN SIGNPOST MATHS NSW 2* • ISBN 9780655709039

31A Repeated subtraction

How many tins can I fill?

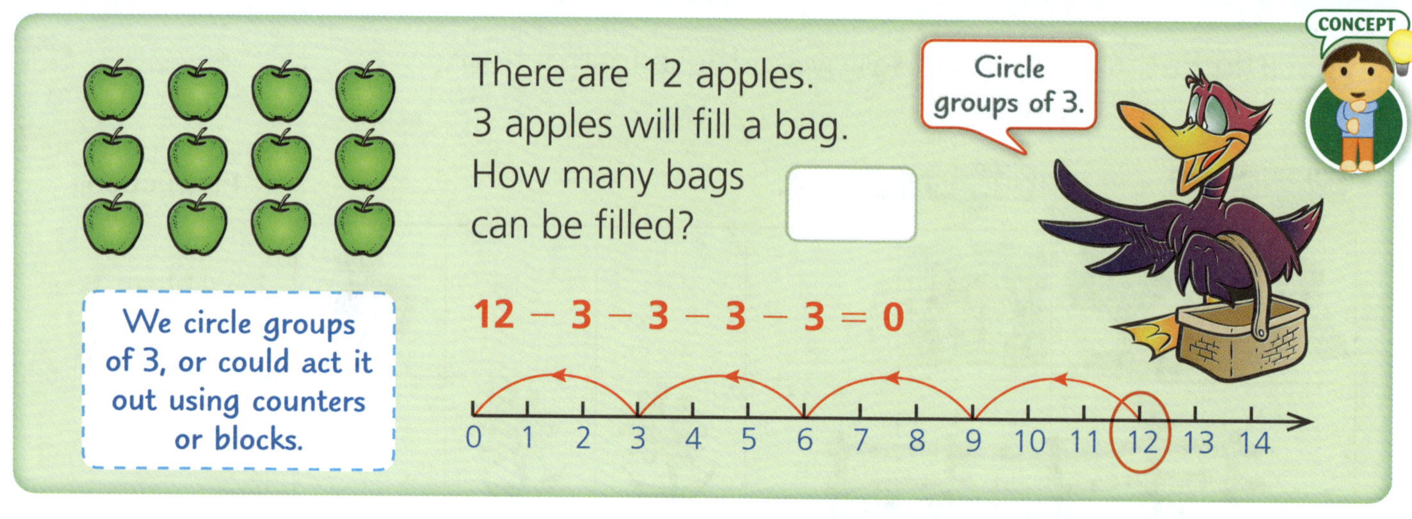

1. We have 12 apples. Write the number of students that can be given:

   a 2 apples ☐ b 4 apples ☐ c 6 apples ☐

2. 

   We have 16 balls.
   How many children can be given:

   a 4 balls? ☐ b 8 balls? ☐
   c 2 balls? ☐ d 1 ball? ☐

3. 

   We have 15 birds.
   How many people can be given:

   a 5 birds? ☐ b 3 birds? ☐
   c 4 birds? ☐ with ☐ birds left over
   d 6 birds? ☐ with ☐ birds left over

4. We have 20 lollies to share.
   How many children can be given:

   a 2 lollies? ☐ b 4 lollies? ☐
   c 5 lollies? ☐ d 10 lollies? ☐
   e 6 lollies? ☐ with ☐ lollies left over

 • *AUSTRALIAN SIGNPOST MATHS NSW 2* • ISBN 9780655709039

31B Fractions of a whole

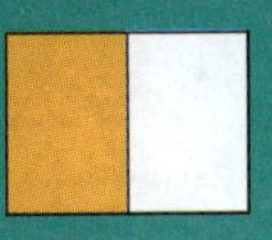

CONCEPT

- One half is one of two equal parts.
- One quarter is one of four equal parts.
- One eighth is one of eight equal parts.

Four quarters make one whole.

Two halves make one whole.

1 What part of the shape is coloured: one half, one quarter or one eighth?

a

b

c

d

e

f

2 Colour part of each shape to match the given fraction.

a one eighth

b one half

c one quarter

INVESTIGATION

- Draw lines to cut this shape into halves. Then draw lines to cut each half into two equal parts (quarters).
- Draw two lines to cut this shape into quarters. Then draw lines to cut each quarter into two equal parts (eighths).

 • *AUSTRALIAN SIGNPOST MATHS NSW 2* • ISBN 9780655709039

Has this been divided into quarters?

1 Match each diagram to a label.

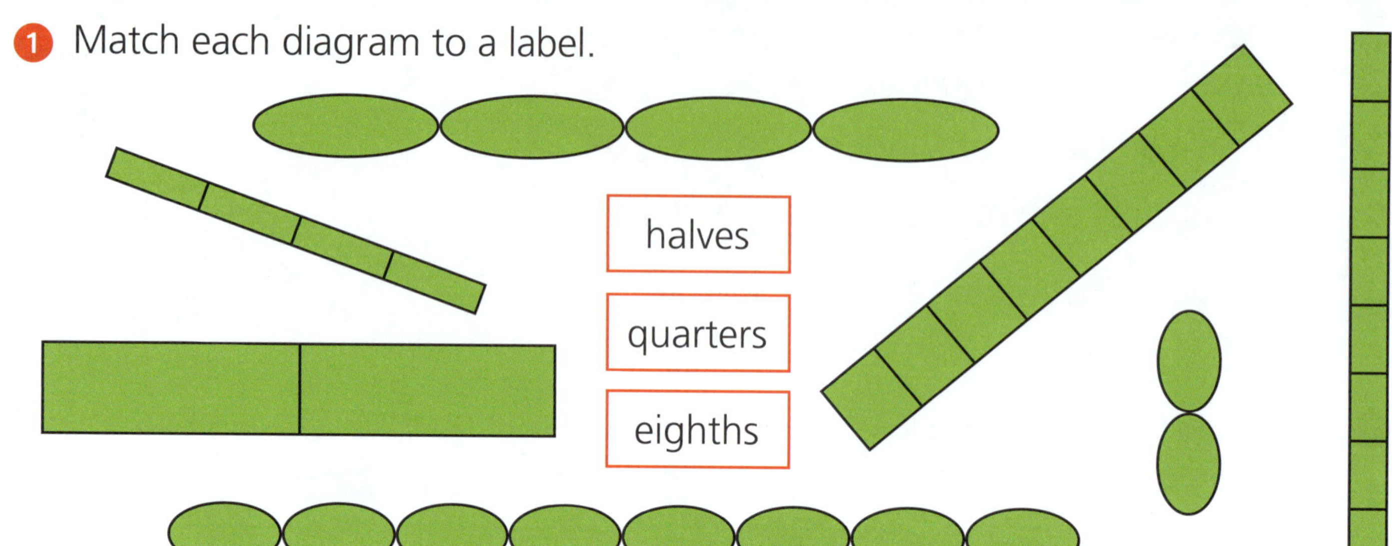

2 Draw lines to cut this paper into halves, then quarters and then eighths.

3 Colour one half.

Colour one quarter.

Colour one eighth.

4 Underline the one that is larger.

a one half or one quarter

b one quarter or one eighth

c one half or one eighth

d one half or one whole

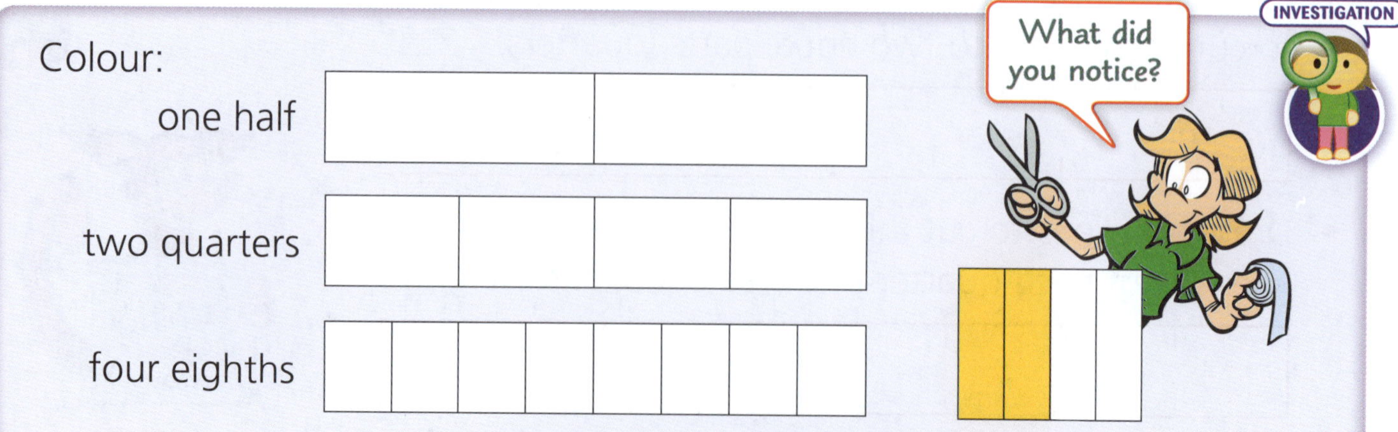

Metres and centimetres

CONCEPT

- The metre is a standard unit of the metric system.
- One metre is usually about the length of your arm span, or two of your steps.
- 1 metre (1 m) = 100 centimetres (100 cm).
  1 cm — This length is 1 centimetre.

ACTIVITY

1. Estimate then measure each length in metres and centimetres.

Length	Estimate	Measurement
length of 5 steps	m	m cm
height of a door	m	m cm
width of a cupboard	m	m cm
width of the classroom	m	m cm

An estimate is a good guess.

m stands for metre.

2. Match each length with the best answer.

width of a door	1 metre	height of yourself
length of a whiteboard	2 metres	length of a car
height of a door	5 metres	width of your desk
width of a window		length of ten steps

3. Measure the following. Show whether each length is less than, about, or more than 1 metre.

	Less than 1 m	About 1 m	More than 1 m
shoulder to toe			
standing jump			
height of your teacher			

Why do we need to use metres and centimetres to measure length?

 • *AUSTRALIAN SIGNPOST MATHS NSW 2* • ISBN 9780655709039

32A Choosing a strategy

1 Match each question with the most appropriate strategy. Explain why you chose each strategy.

doubles	36 + 10	near doubles
	57 − 12	
split strategy	22 + 3	jump strategy
	6 + 7	
adding 10	19 − 3	counting on
	42 − 16	
addition facts	17 + 3	counting back

2 Write the best strategies to solve each problem. Discuss why you chose each strategy.

a 64 − 10 = ☐ *subtracting 10 or using blocks*

b 39 + 4 = ☐

c 9 + 9 = ☐

d 29 + 34 = ☐

e 7 + 8 = ☐

f 64 − 23 = ☐

g 27 + 3 = ☐

h 18 − 9 = ☐

3 Use the jump strategy to answer each of these. Check with a calculator.

a 14 + 69 = ☐ b 83 − 58 = ☐

c 24 + 77 = ☐ d 94 − 69 = ☐

e 73 + 18 = ☐ f 72 − 36 = ☐

32B Choosing a strategy

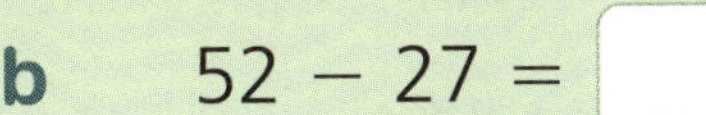

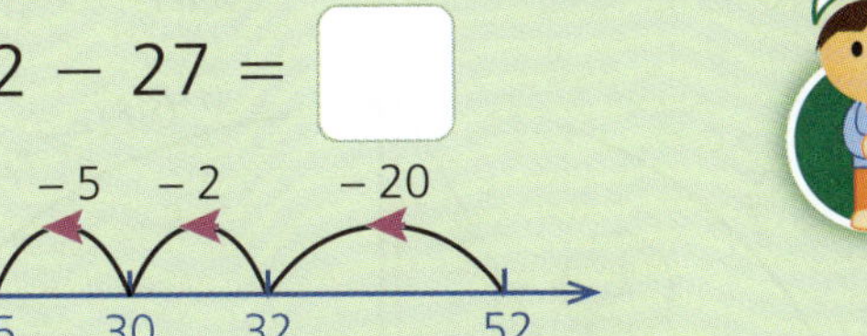

a 49 + 14 = ☐

(40 + 10) + (9 + 4)

= 50 + 13

= **63**

I used the split strategy.
I added the tens, then the ones.

b 52 − 27 = ☐

−5 −2 −20

25 30 32 52

52 − 20 − 2 − 5

= **25**

I used the jump strategy.

1 Use the boxes to show how you found each answer.

a 18 + 9 = ☐

b 56 − 10 = ☐

c 36 + 4 = ☐

d 41 − 4 = ☐

e 28 + 16 = ☐

f 72 + 15 = ☐

 • *AUSTRALIAN SIGNPOST MATHS NSW 2* • ISBN 9780655709039

32C Quarter turns

a quarter turn clockwise
(the way a clock goes)

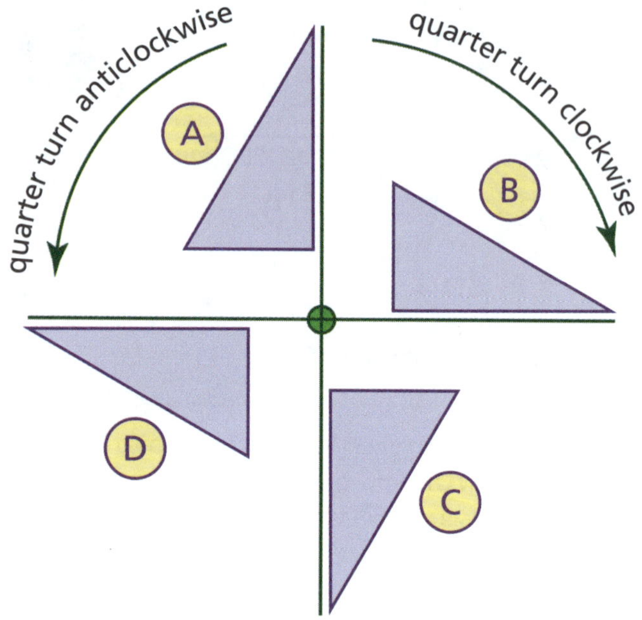

turn to the left

turn to the right

quarter turn anticlockwise

quarter turn clockwise

A B C D

1 Write the letter next to the shape you get if you move shape (A):

a a quarter turn clockwise ☐

b a half turn clockwise ☐

c a quarter turn anticlockwise ☐

d a half turn anticlockwise ☐

e a quarter turn to the right ☐

2 Is a half turn clockwise the same as a half turn anticlockwise? ☐

3 How had triangle (A) changed after it had moved through a half turn?

____

4 The 1st pattern is made using quarter turns. Complete the 2nd pattern.

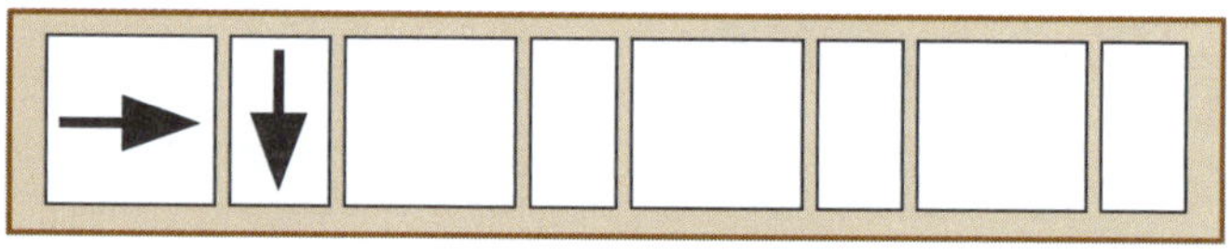

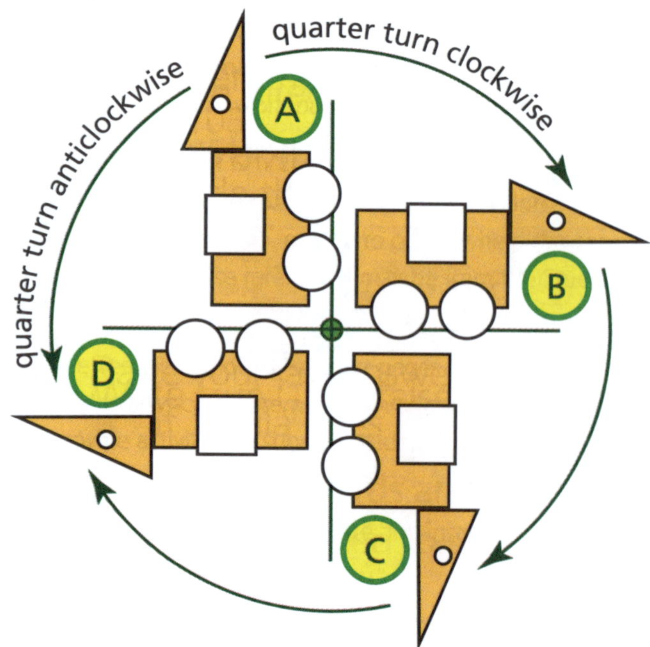

5 Which picture does (B) become if you turn it:

a a quarter turn clockwise? ☐

b a half turn clockwise? ☐

c a quarter turn anticlockwise? ☐

d a half turn anticlockwise? ☐

6 If you turn (A) through a full turn, which picture does it become? ☐

7 If you turn a shape, does it stay the same shape and size? ☐

32D Half and quarter turns

turn to the left – anticlockwise

turn to the right – clockwise

1 Write a "H" for the half turns and a "Q" for the quarter turns.

a 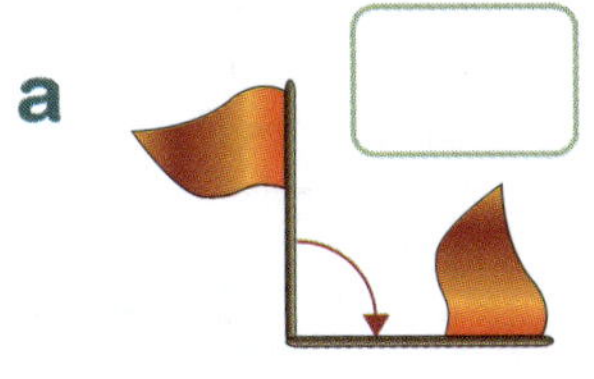
b 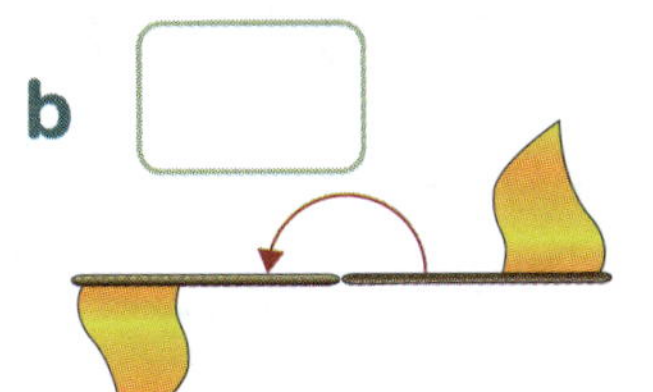
c 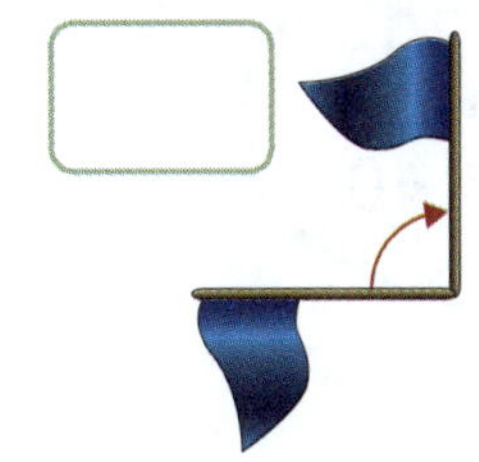
d 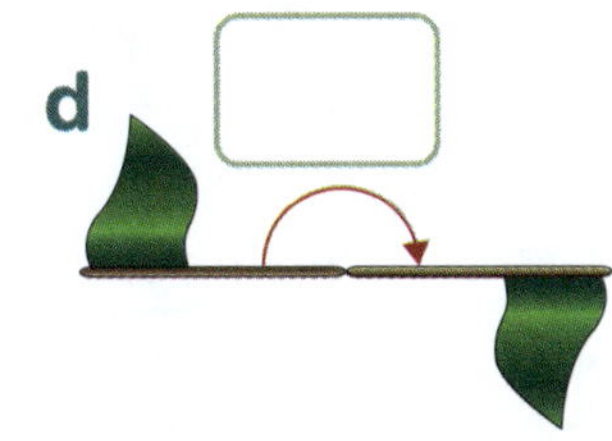

2 Continue the **quarter turn** pattern to finish each design.

a 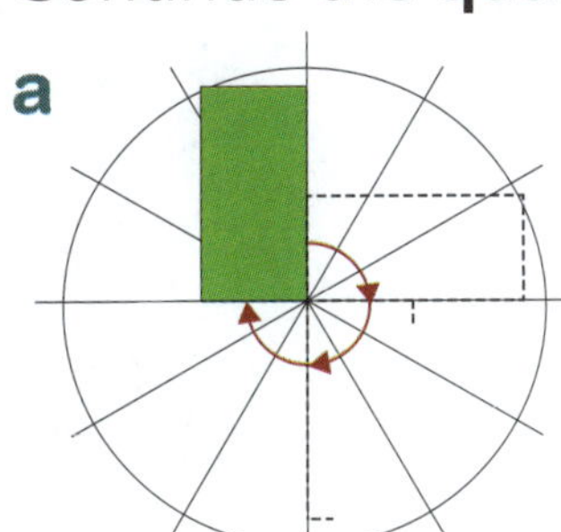
b 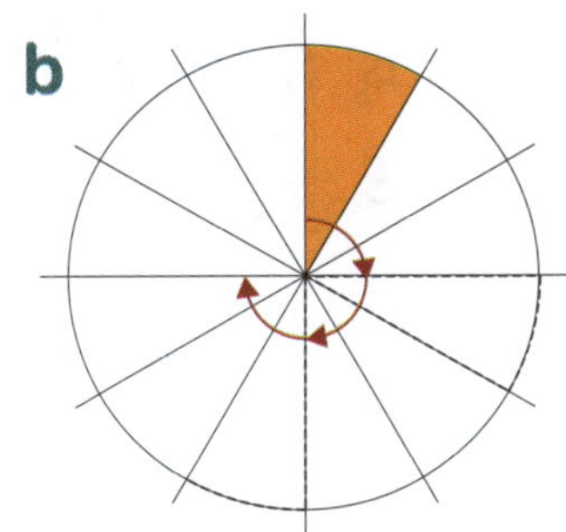
c 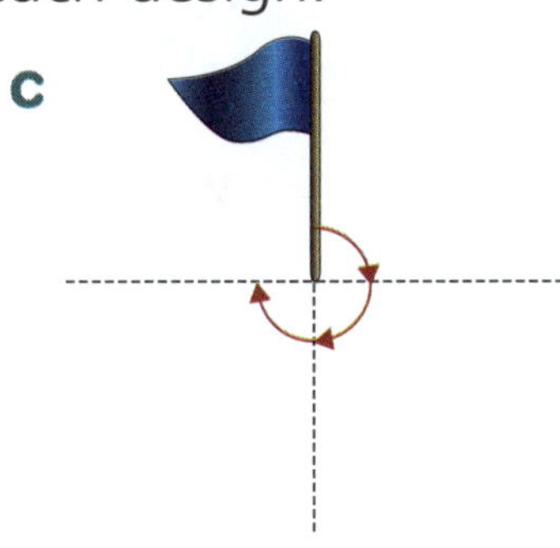

3 Continue the **half turn** pattern to finish each design.

a 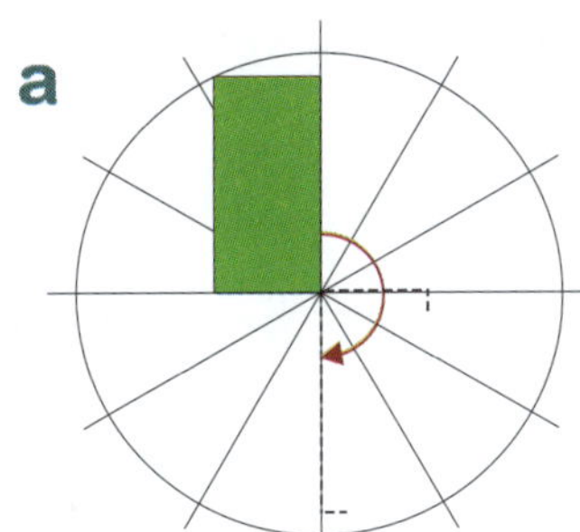
b 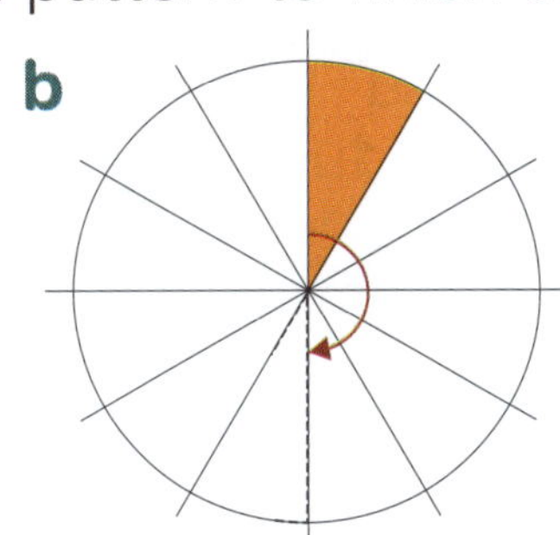
c 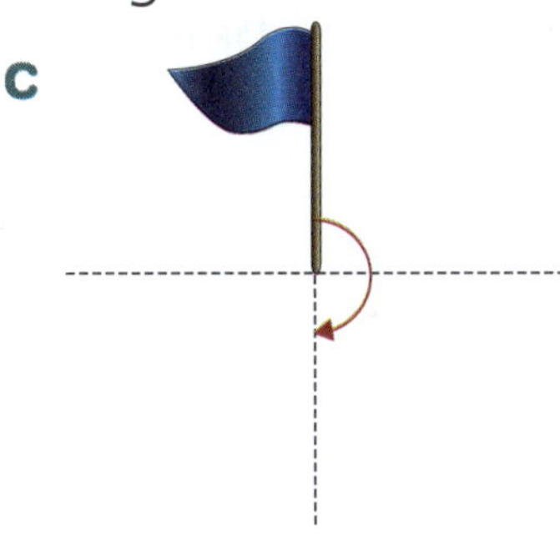

4 Describe each turn in Question 1 as either clockwise or anticlockwise.

a

b

c

d

INVESTIGATION

Show where each minute hand will be after the turn.

Start

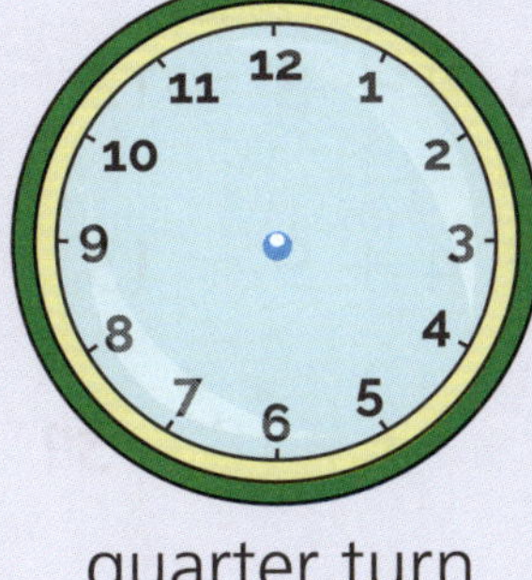

quarter turn clockwise

half turn clockwise

full turn clockwise

 • *AUSTRALIAN SIGNPOST MATHS NSW 2* • ISBN 9780655709039

33A Related problems

12 + 8 = 20
so
12 + 18 = 30

9 + 21 = 30
so
19 + 21 = 40

CONCEPT

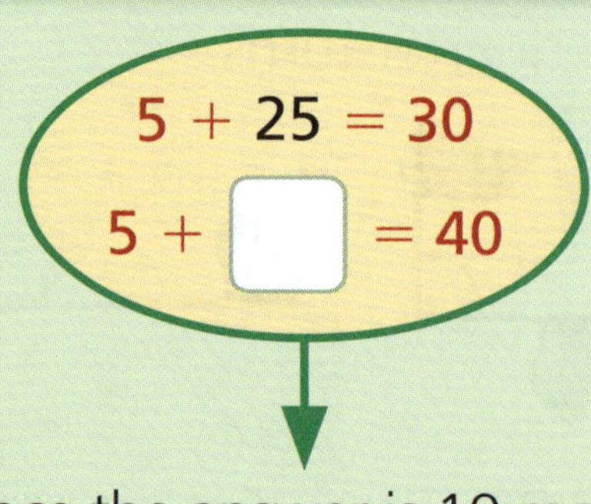

Since the answer is 10 more, we must have added 10 more.

5 + 35 = 40

The more we add, the larger the answer will be.

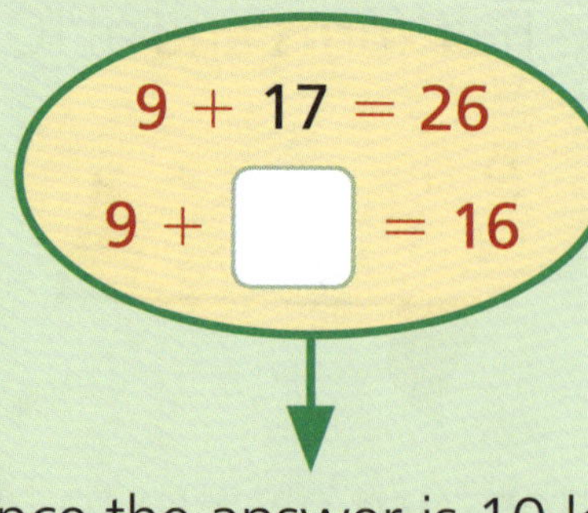

Since the answer is 10 less, we must have added 10 less.

9 + 7 = 16

The less we add, the smaller the answer will be.

1. Use the first number sentence to work out the other two answers.

a 4 + 8 = 12 so 4 + 18 = ☐ and 4 + 28 = ☐

b 7 + 6 = 13 so 7 + 16 = ☐ and 7 + 26 = ☐

c 8 + 7 = 15 so 8 + 17 = ☐ and 8 + 27 = ☐

d 4 + 6 = 10 so 4 + 16 = ☐ and 4 + 26 = ☐

e 6 + 7 = 13 so 6 + 17 = ☐ and 6 + 27 = ☐

f 17 + 9 = 26 so 17 + 19 = ☐ and 17 + 29 = ☐

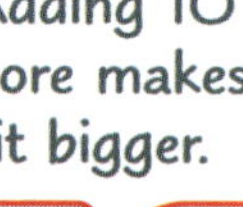

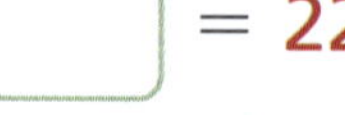

2. Use the first number sentence to work out the other answer.

a 9 + 3 = 12 so 9 + ☐ = 22

b 4 + 8 = 12 so 4 + ☐ = 22

c 8 + 6 = 14 so 8 + ☐ = 24

d 4 + 9 = 13 so 4 + ☐ = 23

e 6 + 19 = 25 so 6 + ☐ = 35

f 13 + 18 = 31 so 13 + ☐ = 21

g 5 + 19 = 24 so 5 + ☐ = 14

h 8 + 19 = 27 so 8 + ☐ = 17

Inverse strategy, subtraction

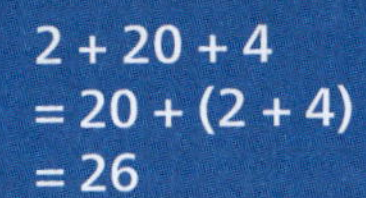

CONCEPT

74 – 48

Use a number line to go from the smaller number to the larger one.

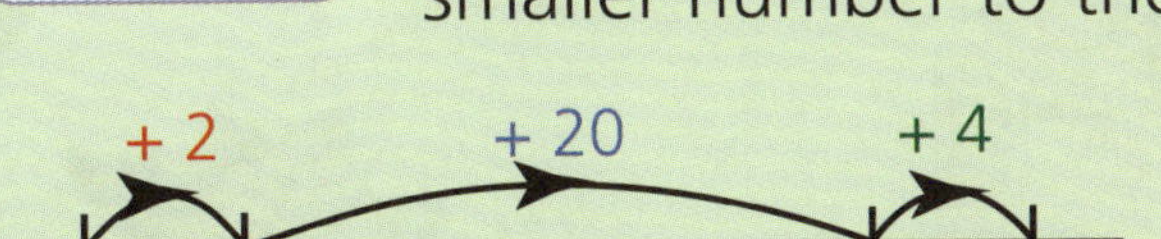

Start at **48**, add 2 to make 50,
then add 20 to make 70,
then add 4 to make **74**.

Total the bits added on.

74 – 48
= 2 + 20 + 4
= 26

This is called the shopkeeper's method.

1 Find the total of these numbers.

a 3, 10 and 2 ☐ **b** 6, 30 and 1 ☐ **c** 4, 60 and 3 ☐

2 Use the inverse strategy to complete each question.

a 62 – 17 Start at **17**, add 3 (20) then add 40 (60) then add 2 (62).

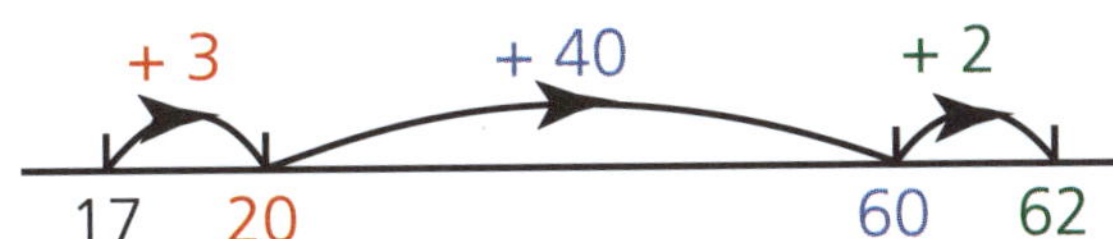

3 + 40 + 2 = ☐

b 51 – 24 Start at **24**, add 6 (20) then add 30 (50) then add 1 (51).

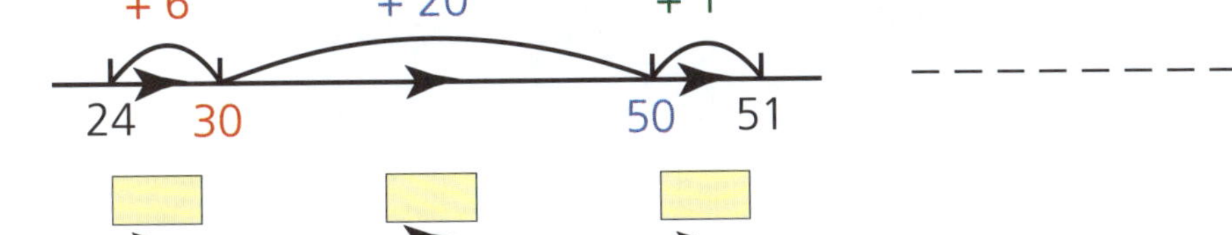

__________ ☐

c 93 – 26

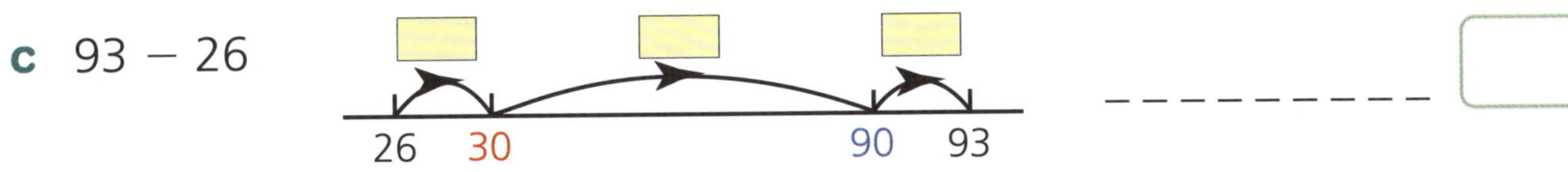

__________ ☐

d 86 – 49

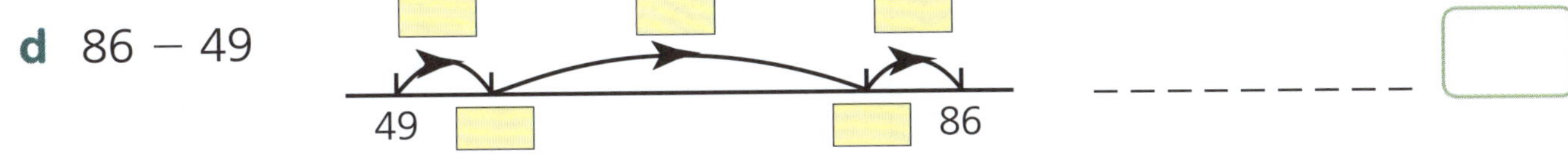

__________ ☐

3 Use the inverse strategy to find each answer.

a 64 – 25 Add ____ , then add ____ , then add ____. ______ ☐

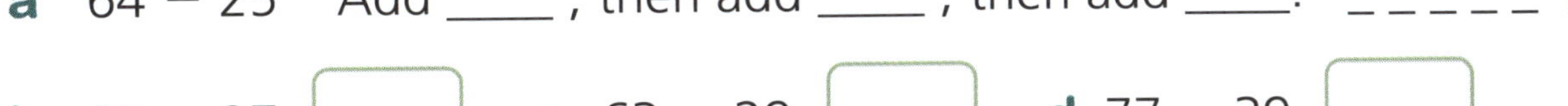

b 85 – 27 ☐ **c** 63 – 29 ☐ **d** 77 – 39 ☐

e 47 – 18 ☐ **f** 82 – 56 ☐ **g** 94 – 67 ☐

Use addition.

 • *AUSTRALIAN SIGNPOST MATHS NSW 2* • ISBN 9780655709039

33C Describing 3D objects

Terms:
cube, prism, cone, pyramid, cylinder, sphere

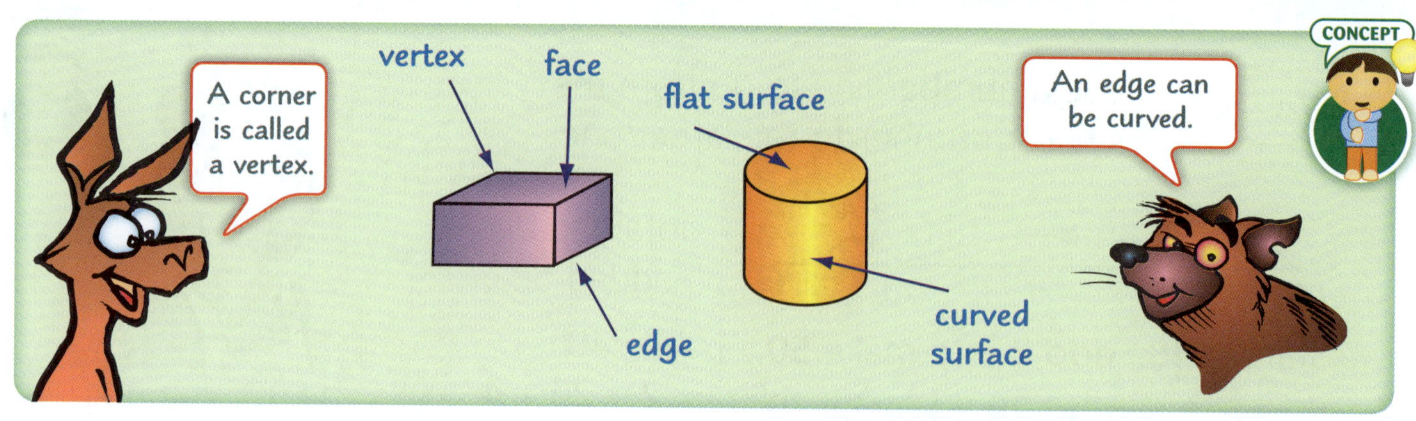

1 Find objects like these and complete the tables.

Object	Number of faces	Number of edges	Number of vertices	Number of curved surfaces
A				
B				

A face is a flat surface with straight sides.

Vertices are corners.

Object	Number of curved surfaces	Number of flat surfaces
C		
D		
E		

Name these objects.

A

B

C

D

E

2 What 2D shapes are on the pyramid in object B?

33D 3D objects

one vertex, two vertices

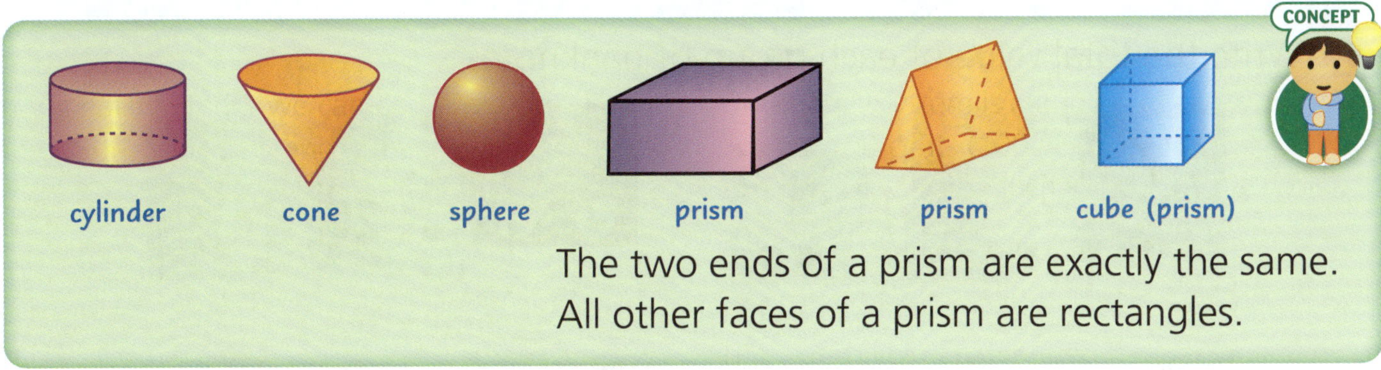

The two ends of a prism are exactly the same.
All other faces of a prism are rectangles.

1 Write the name of each 3D object.

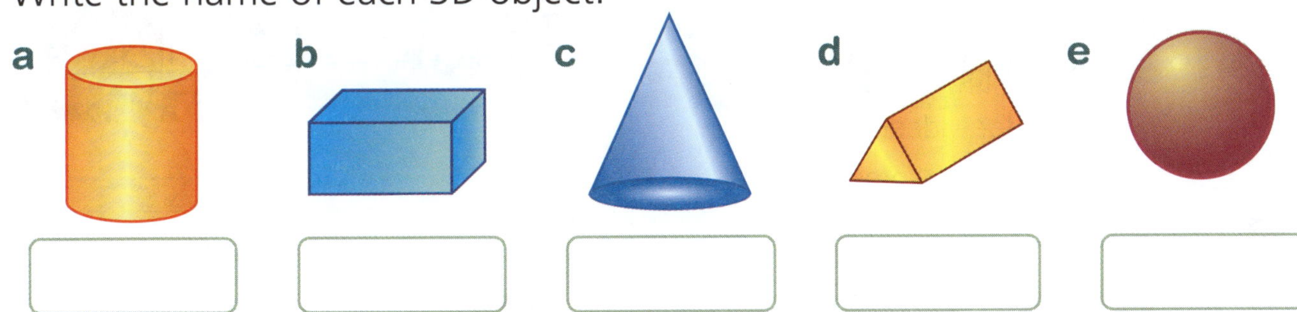

2

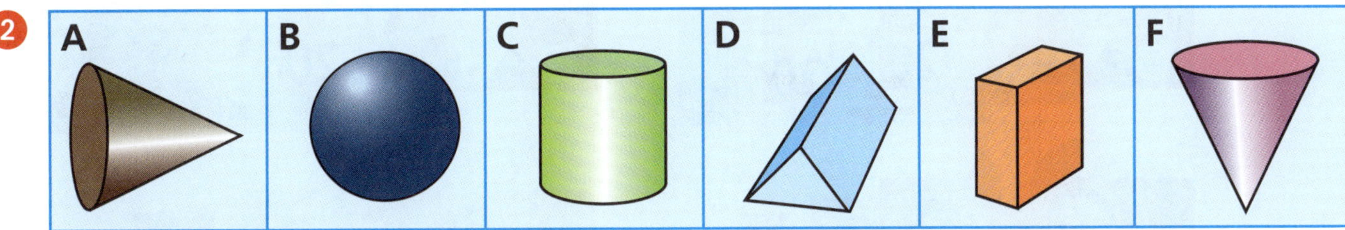

Which 3D objects have:

a curved surfaces?

b only flat surfaces?

c both curved and flat surfaces?

3 A **face** is a flat surface that has only straight sides.

 • *AUSTRALIAN SIGNPOST MATHS NSW 2* • ISBN 9780655709039

Money

1 Write the total value of each group of banknotes.

a

b

c

d

2 Circle the larger amount.

a

b

c

3 Write the amounts in order from smallest to largest.

a

b

c

$85

34B Possible outcomes

1 **a** Is it possible that it will rain tomorrow?

b Is it possible to roll a zero on a normal die?

c Is it possible that you will live to be 100?

2 **a** What are the possible outcomes if I play a board game?

b What kinds of weather are possible for tomorrow?

3

red
red
red
blue
blue
green

a List all possible outcomes for this spinner.

b Which colour is most likely to be spun?

c Which colour is least likely to be spun?

d Is it possible to spin yellow?

4 List all possible outcomes when a die is rolled.

Is a 6 as likely as a 1?

INVESTIGATION

- Toss a coin until you have tossed three heads. Have a friend count the number of tosses needed. ___ tosses
- Guess how many times it will take to toss three heads next time. ___ Repeat the experiment. It took ___ tosses.

Tally marks

Tallies are usually placed in groups of 5. The fifth mark is usually drawn across the other four marks.

卌 卌 = 10

1 Make up a question and write 4 answers. Ask 15 students in your class to make a choice.

Write your question here.

4 possible choices:

2 Show the answers given by 15 students.
Put your 4 possible choices in the left column.
Trace one line in black for each student.

3 Make a picture graph on the right.
Put your 4 possible choices in the spaces at the bottom of the graph. Use the information in the table to colour a face for each student who chose that answer.

 • *AUSTRALIAN SIGNPOST MATHS NSW 2* • ISBN 9780655709039

34D Comparing objects

Objects can have length, area, volume and mass.

CONCEPT

An estimate is your best guess.

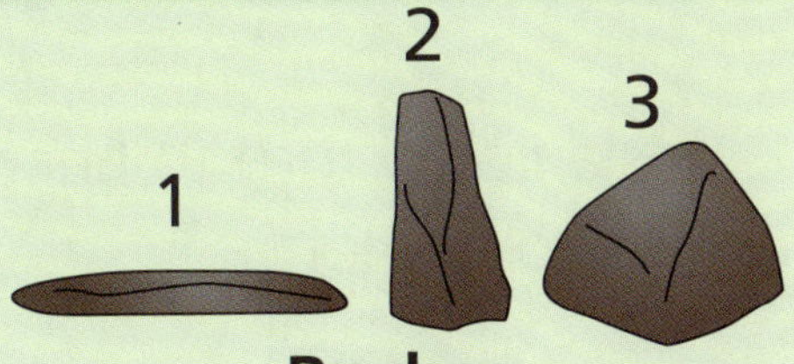

Rocks: 1, 2, 3

You can use ones blocks to measure lengths.

1. Estimate which of these rocks is:

a the tallest ☐ b the shortest ☐ c the heaviest ☐
d the lightest ☐ e the widest ☐ f the narrowest ☐
g the one that takes up the most space (has the largest volume) ☐
h the one that takes up the least space (has the smallest volume) ☐

Containers: 1, 2, 3

These are filled with water.

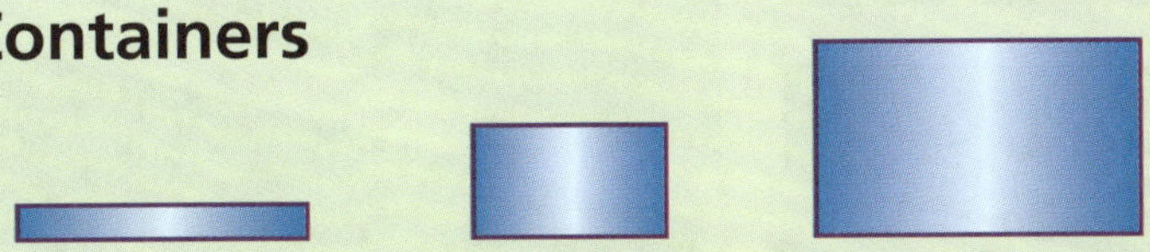

2. Estimate which of these containers:

a holds the most water ☐ b is the lightest ☐
c is the widest ☐ d holds the least water ☐
e is the heaviest ☐ f is the narrowest ☐
g has the most outside area ☐

ACTIVITY

3. Find three objects and draw them. Compare them. Draw lines to match each object with its properties.

Tallest | Shortest | Biggest area | Smallest area | Heaviest | Lightest

Use your hand as a unit of area.

Use balance scales to find the heaviest.

 • *AUSTRALIAN SIGNPOST MATHS NSW 2* • ISBN 9780655709039

35A Giving directions

Turn right means turn clockwise.

Turn left means turn anticlockwise.

1 a Use this map and the **words** above to explain how to go from B to F.

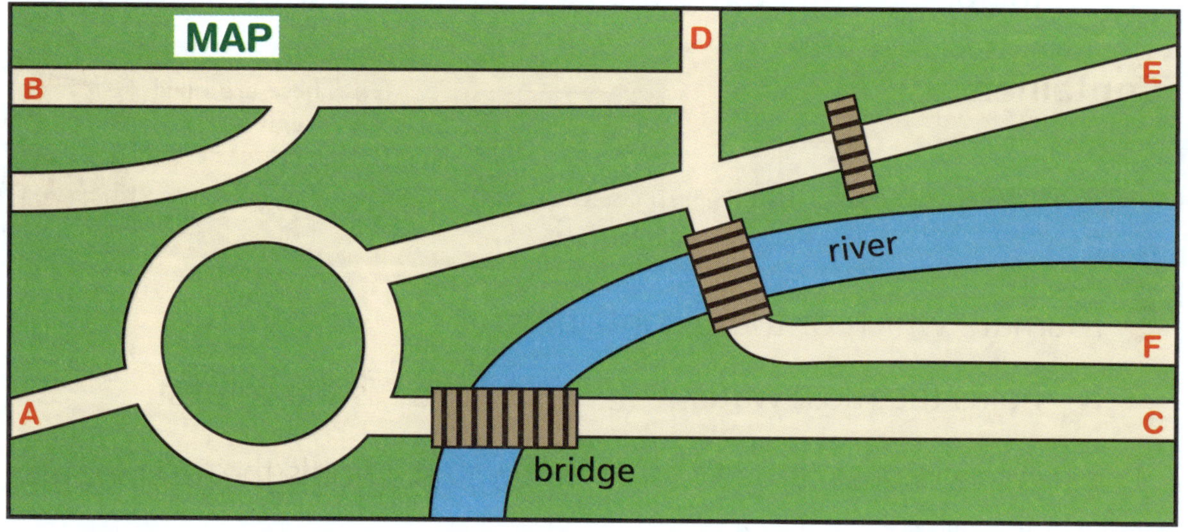

b Give directions to go from A to another letter on the map.

Work with a partner and take turns.

2

a If Jamila goes forwards 3 squares, she reaches ☐.

b What forwards and backwards directions does she need to spell the word "CAKE"? Discuss.

3 Use directions to explain how to get to the principal's office.

35B More shapes (extension)

These shapes are **quadrilaterals**. They have 4 straight sides and 4 vertices. **Parallel lines** go in the same direction. They are always the same distance apart.

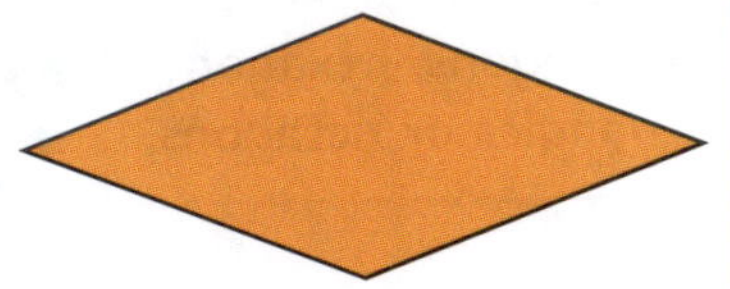

A **rhombus** has all sides equal. A diamond is a rhombus.

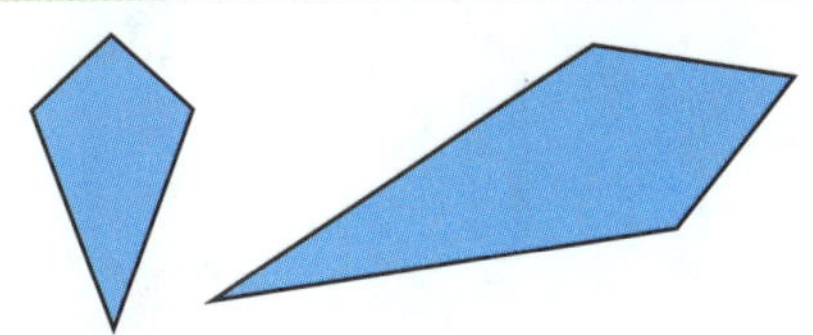

A **kite** has two pairs of equal sides. Toy kites can fly.

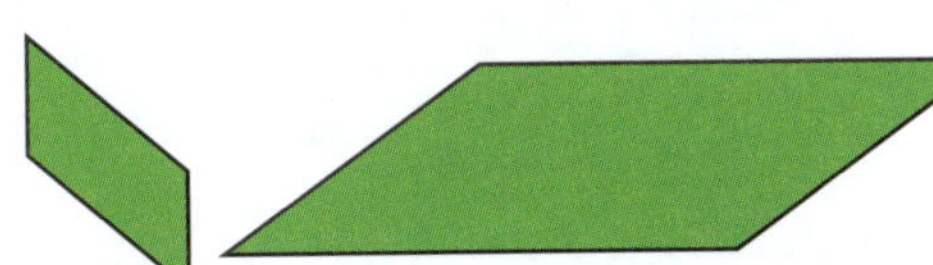

A **parallelogram** has opposite sides equal and parallel.

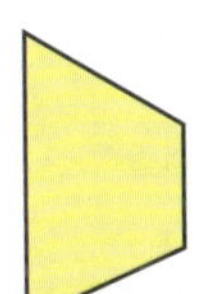
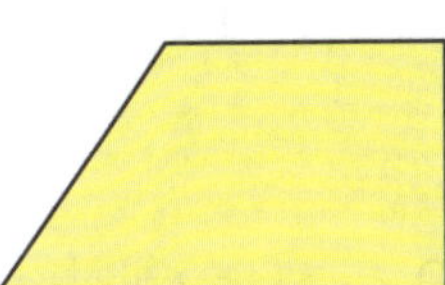

A **trapezium** has one pair of parallel sides.

1 Put a letter on each shape. Write:

R for rhombus K for kite

P for parallelogram T for trapezium.

Squares and rectangles are also quadrilaterals.

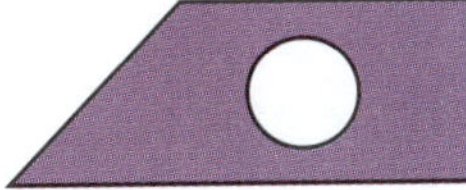
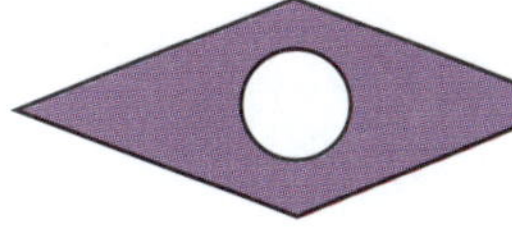

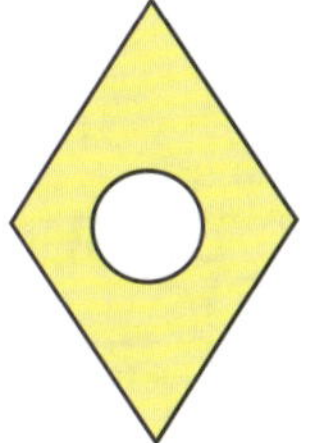
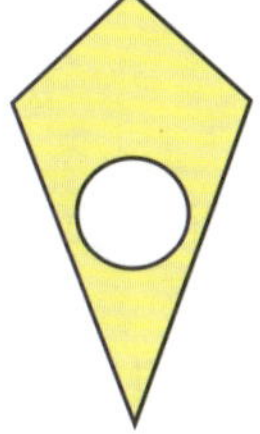
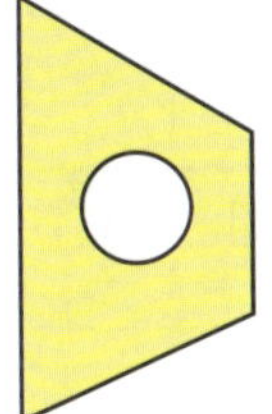
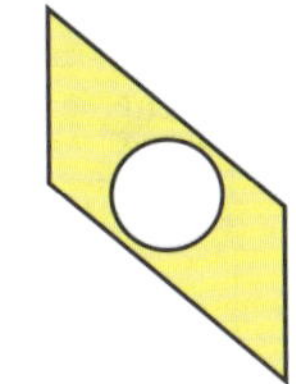

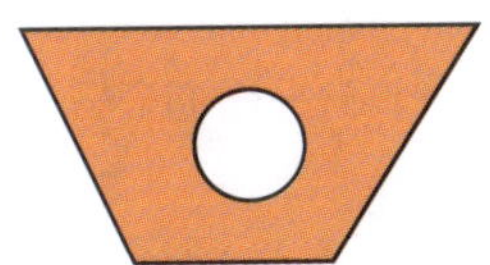
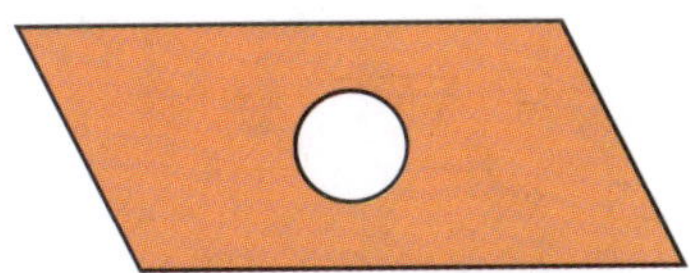
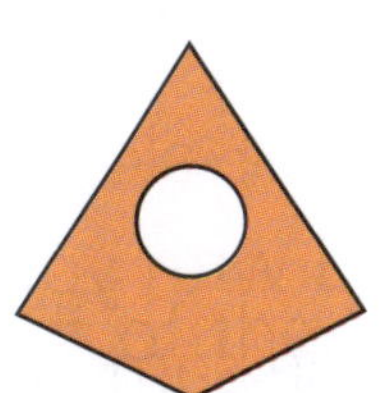
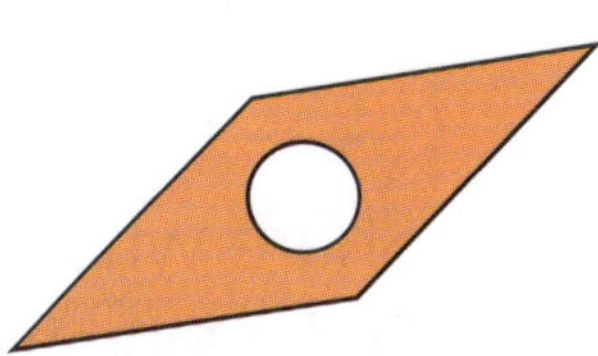

 ISBN 9780655709039

35C Problem solving with addition

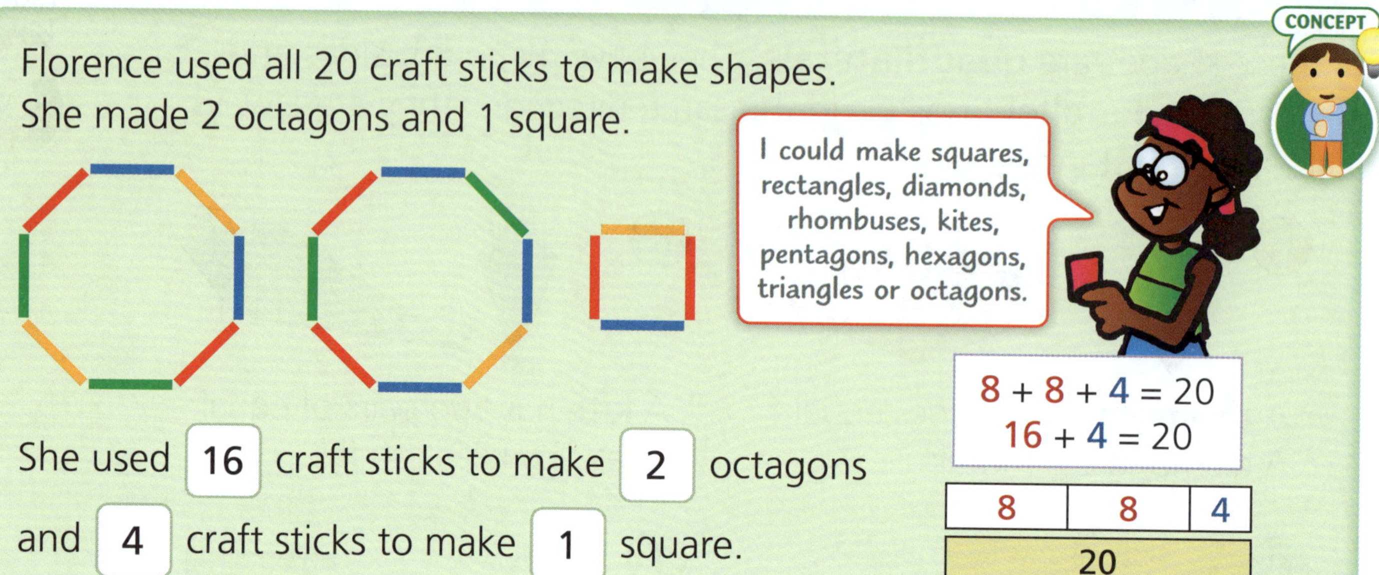

Florence used all 20 craft sticks to make shapes.
She made 2 octagons and 1 square.

She used 16 craft sticks to make 2 octagons and 4 craft sticks to make 1 square.

1 **a** Use 20 crafts sticks to see what other combinations of shapes you can make. Try to use all of the craft sticks. Draw and write about what you did.

b Share what you did with your class. Did everyone get the same answer?

 • *AUSTRALIAN SIGNPOST MATHS NSW 2* • ISBN 9780655709039

Problem solving with groups

2, 4, 6, 8, 10, 12.
6 groups of 2
makes 12.

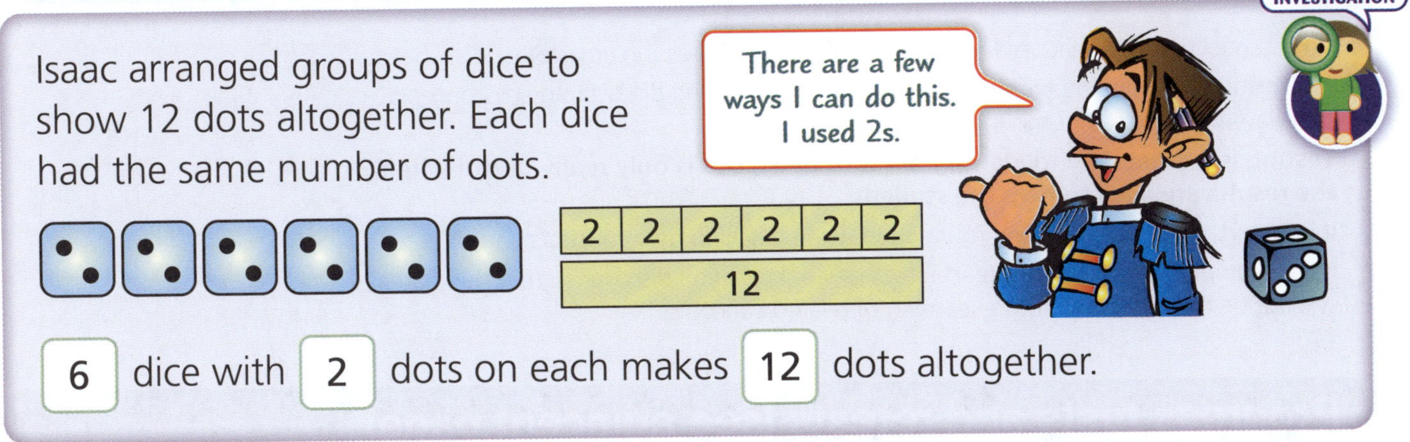

1 **a** Show two other ways Isaac could arrange the dice so that 12 dots are shown in each box. Each box must have the same number on each die.

b Write about what you did in one of your boxes.

2

5 + 5 + 5 = 15
3 pentagons would have 15 sides.

Repeat the same shape in each box to match the number of sides given. Write the shape you used.

a 18 **b** 16 **c** 24

Identifying and addressing areas of need

An essential part of a teacher's role is identifying and addressing areas of student need. This includes recognising areas where memory is fading and discovering any concepts that have been missed or misunderstood.

Testing is a great way to identify areas of need, but is only really useful when the results are used to help the student.

It is important to build a strong foundation when teaching new concepts and skills.

It is also important to revise/re-teach areas of weakness you discover so that these areas will not be barriers to the future learning of related concepts.

Progress tests and retests (see adjacent page)

Progress tests 1 to 5 are found on pages 147–169 of the online Teacher Resource.

After each test, notes and answers are supplied.

Progress test questions are cross-referenced to appropriate Student Book pages.

Progress retests 1 to 5 are found on pages 172–194 of the online Teacher Resource.

The remediation records pages are used to provide a record of each student's progress.

These are found on pages 145–146 and 170–171 of the online Teacher Resource.

For each error recorded, the question should be discussed, and using the Student Book cross-reference provided, practice should occur. Retesting should follow using the progress retests.

Summary

1 Test recent work.

2 Enter any mistakes in the Remediation records.

3 Use this record to direct your revision/re-teaching.

4 Retest using the matching retest questions to ensure understanding.

Teaching and learning

Successfully teaching content and skills is a complex process.

A **good textbook** is an important tool alongside **effective teaching and planning**.

Knowledge, understanding and skills must be embedded in the student's mind so that recall continues with time. This will be done using:

(1) instruction (2) practice (3) drill (4) review.

Instruction involves explicit explanation, investigation and the use of good educational resources.

Practice forms neural pathways within the brain.

Drill strengthens neural pathways. The stronger the pathways become, the longer the understanding or knowledge is retained. 'Overlearning' prolongs recall.

Review revitalises weakened neural pathways.

Progress test 2

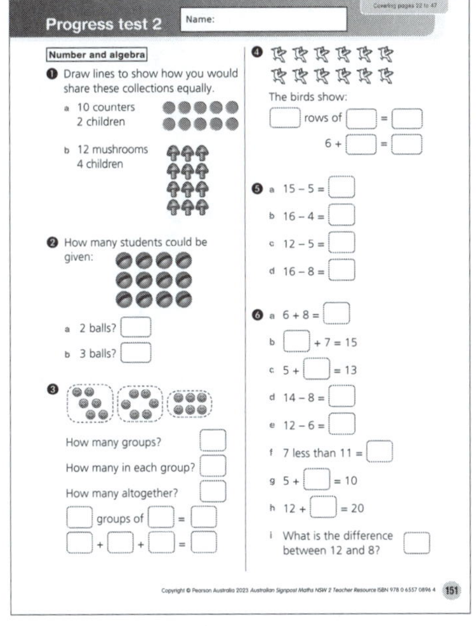
Covering pages 22 to 47

Progress test 2

Name:

Number and algebra

1 Draw lines to show how you would share these collections equally.
a 10 counters, 2 children
b 12 mushrooms, 4 children

2 How many students could be given:
a 2 balls?
b 3 balls?

3 How many groups?
How many in each group?
How many altogether?
☐ groups of ☐ = ☐
☐ + ☐ + ☐ = ☐

4 The birds show:
☐ rows of ☐ = ☐
6 + ☐ = ☐

5 a 15 − 5 =
b 16 − 4 =
c 12 − 5 =
d 16 − 8 =

6 a 6 + 8 =
b ☐ + 7 = 15
c 5 + ☐ = 13
d 14 − 8 =
e 12 − 6 =
f 7 less than 11 =
g 5 + ☐ = 10
h 12 + ☐ = 20
i What is the difference between 12 and 8?

Copyright © Pearson Australia 2023 Australian Signpost Maths NSW 2 Teacher Resource ISBN 978 0 6557 0896 4 151

Progress retest 2

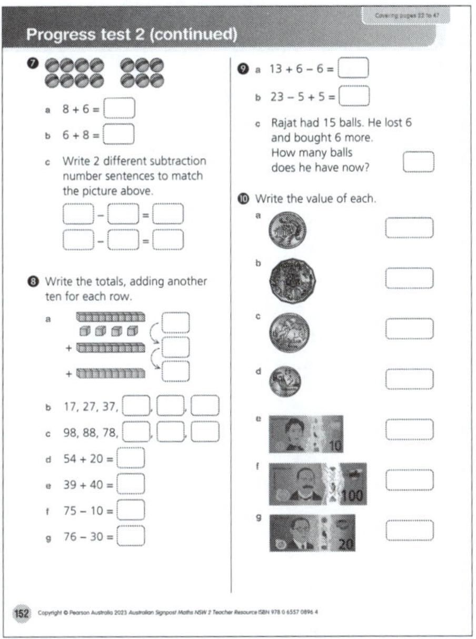
Covering pages 22 to 47

Progress test 2 (continued)

7 a 8 + 6 =
b 6 + 8 =
c Write 2 different subtraction number sentences to match the picture above.
☐ − ☐ = ☐
☐ − ☐ = ☐

8 Write the totals, adding another ten for each row.
a
b 17, 27, 37, ☐, ☐, ☐
c 98, 88, 78, ☐, ☐, ☐
d 54 + 20 =
e 39 + 40 =
f 75 − 10 =
g 76 − 30 =

9 a 13 + 6 − 6 =
b 23 − 5 + 5 =
c Rajat had 15 balls. He lost 6 and bought 6 more. How many balls does he have now?

10 Write the value of each.
a b c d e f g

152 Copyright © Pearson Australia 2023 Australian Signpost Maths NSW 2 Teacher Resource ISBN 978 0 6557 0896 4

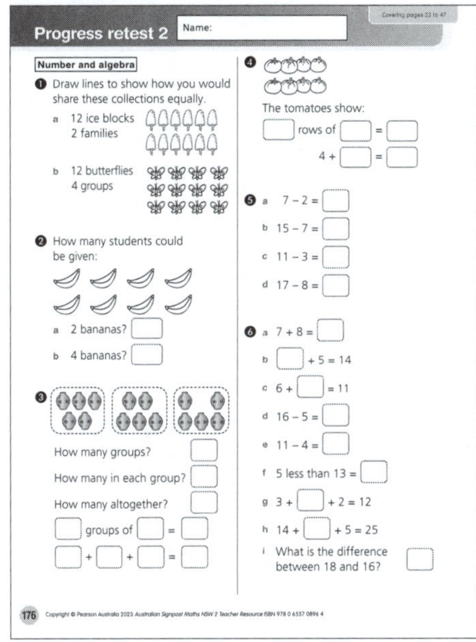
Covering pages 22 to 47

Progress retest 2

Name:

Number and algebra

1 Draw lines to show how you would share these collections equally.
a 12 ice blocks, 2 families
b 12 butterflies, 4 groups

2 How many students could be given:
a 2 bananas?
b 4 bananas?

3 How many groups?
How many in each group?
How many altogether?
☐ groups of ☐ = ☐
☐ + ☐ + ☐ = ☐

4 The tomatoes show:
☐ rows of ☐ = ☐
4 + ☐ = ☐

5 a 7 − 2 =
b 15 − 7 =
c 11 − 3 =
d 17 − 8 =

6 a 7 + 8 =
b ☐ + 5 = 14
c 6 + ☐ = 11
d 16 − 5 =
e 11 − 4 =
f 5 less than 13 =
g 3 + ☐ + 2 = 12
h 14 + ☐ + 5 = 25
i What is the difference between 18 and 16?

176 Copyright © Pearson Australia 2023 Australian Signpost Maths NSW 2 Teacher Resource ISBN 978 0 6557 0896 4

Notes and answers for Progress test 2

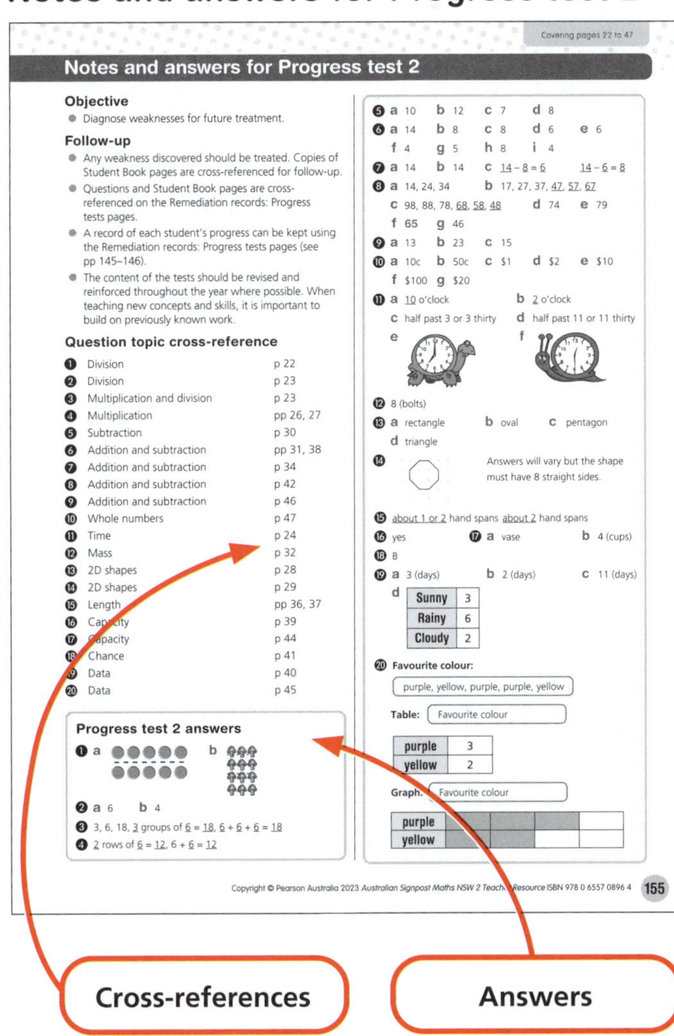
Covering pages 22 to 47

Notes and answers for Progress test 2

Objective
- Diagnose weaknesses for future treatment.

Follow-up
- Any weakness discovered should be treated. Copies of Student Book pages are cross-referenced for follow-up.
- Questions and Student Book pages are cross-referenced on the Remediation records: Progress tests pages.
- A record of each student's progress can be kept using the Remediation records: Progress tests pages (see pp 145–146).
- The content of the tests should be revised and reinforced throughout the year where possible. When teaching new concepts and skills, it is important to build on previously known work.

Question topic cross-reference

1	Division	p 22
2	Division	p 23
3	Multiplication and division	p 23
4	Multiplication	pp 26, 27
5	Subtraction	p 30
6	Addition and subtraction	pp 31, 38
7	Addition and subtraction	p 34
8	Addition and subtraction	p 42
9	Addition and subtraction	p 46
10	Whole numbers	p 47
11	Time	p 24
12	Mass	p 32
13	2D shapes	p 28
14	2D shapes	p 29
15	Length	pp 36, 37
16	Capacity	p 39
17	Capacity	p 44
18	Chance	p 41
19	Data	p 40
20	Data	p 45

Progress test 2 answers

1 a b
2 a 6 b 4
3 3, 6, 18, 3 groups of 6 = 18, 6 + 6 + 6 = 18
4 2 rows of 6 = 12, 6 + 6 = 12
5 a 10 b 12 c 7 d 8
6 a 14 b 8 c 8 d 6 e 6 f 4 g 5 h 8 i 4
7 a 14 b 14 c 14 − 8 = 6, 14 − 6 = 8
8 a 14, 24, 34 b 17, 27, 37, 47, 57, 67 c 98, 88, 78, 68, 58, 48 d 74 e 79 f 65 g 46
9 a 13 b 23 c 15
10 a 10c b 50c c $1 d $2 e $10 f $100 g $20
11 a 10 o'clock b 2 o'clock c half past 3 or 3 thirty d half past 11 or 11 thirty e f
12 8 (bolts)
13 a rectangle b oval c pentagon d triangle
14 Answers will vary but the shape must have 8 straight sides.
15 about 1 or 2 hand spans about 2 hand spans
16 yes 17 a vase b 4 (cups)
18 B
19 a 3 (days) b 2 (days) c 11 (days)
d

Sunny	3
Rainy	6
Cloudy	2

20 Favourite colour:
purple, yellow, purple, purple, yellow
Table: Favourite colour

purple	3
yellow	2

Graph: Favourite colour
purple
yellow

Copyright © Pearson Australia 2023 Australian Signpost Maths NSW 2 Teacher Resource ISBN 978 0 6557 0896 4 155

Remediation records: Progress tests

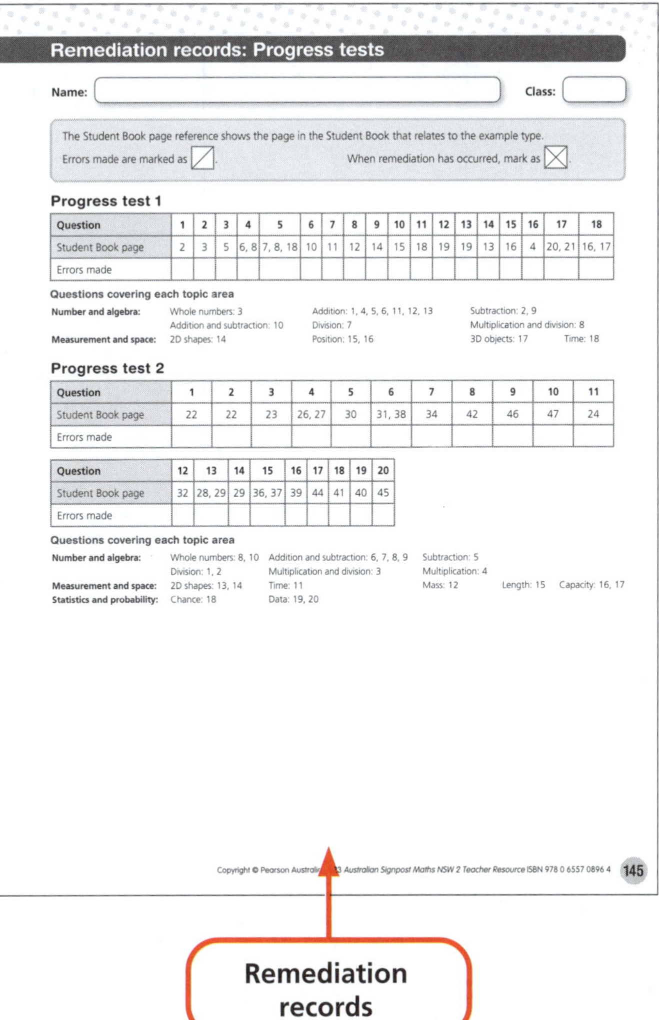

Remediation records: Progress tests

Name: Class:

The Student Book page reference shows the page in the Student Book that relates to the example type.
Errors made are marked as ☐. When remediation has occurred, mark as ☒.

Progress test 1

Question	1	2	3	4	5	6	7	8	9	10	11	12	13	14	15	16	17	18
Student Book page	2	3	5	6, 8	7, 8, 18	10	11	12	14	15	18	19	19	13	16	4	20, 21	16, 17
Errors made																		

Questions covering each topic area
Number and algebra: Whole numbers: 3; Addition: 1, 4, 5, 6, 11, 12, 13; Subtraction: 2, 9; Addition and subtraction: 10; Division: 7; Multiplication and division: 8
Measurement and space: 2D shapes: 14; Position: 15, 16; 3D objects: 17; Time: 18

Progress test 2

Question	1	2	3	4	5	6	7	8	9	10	11
Student Book page	22	22	23	26, 27	30	31, 38	34	42	46	47	24
Errors made											

Question	12	13	14	15	16	17	18	19	20
Student Book page	32	28, 29	29	36, 37	39	44	41	40	45
Errors made									

Questions covering each topic area
Number and algebra: Whole numbers: 8, 10; Addition and subtraction: 6, 7, 8, 9; Subtraction: 5; Division: 1, 2; Multiplication and division: 3; Multiplication: 4
Measurement and space: 2D shapes: 13, 14; Time: 11; Mass: 12; Length: 15; Capacity: 16, 17
Statistics and probability: Chance: 18; Data: 19, 20

Copyright © Pearson Australia 2023 Australian Signpost Maths NSW 2 Teacher Resource ISBN 978 0 6557 0896 4 145

Cross-references

Answers

Remediation records

BLM 1 Addition facts to 20

A	1 + 1		2 + 4		5 + 1		2 + 3		4 + 5	
B	3 + 4		0 + 4		3 + 1		1 + 6		2 + 5	
C	0 + 7		7 + 1		1 + 3		2 + 6		3 + 6	
D	7 + 0		9 + 3		2 + 7		3 + 5		6 + 5	
E	4 + 1		8 + 3		0 + 5		9 + 0		9 + 4	
F	2 + 2		3 + 7		6 + 3		6 + 0		5 + 2	
G	5 + 3		3 + 2		2 + 8		4 + 2		1 + 7	
H	5 + 5		8 + 2		6 + 4		4 + 4		7 + 2	
I	4 + 3		5 + 4		7 + 3		6 + 2		6 + 6	
J	1 + 8		2 + 1		6 + 1		3 + 3		4 + 6	
K	7 + 7		9 + 1		5 + 6		1 + 4		9 + 9	
L	8 + 1		1 + 2		4 + 8		6 + 7		9 + 8	
M	1 + 5		0 + 9		8 + 8		8 + 0		1 + 9	
N	9 + 2		8 + 4		4 + 7		7 + 9		9 + 6	
O	7 + 6		2 + 9		7 + 4		3 + 8		5 + 7	
P	3 + 9		4 + 0		9 + 7		6 + 8		5 + 9	
Q	8 + 6		4 + 9		8 + 7		5 + 8		7 + 5	
R	7 + 8		8 + 9		8 + 5		6 + 9		9 + 5	

BLM 2 Subtraction facts to 10

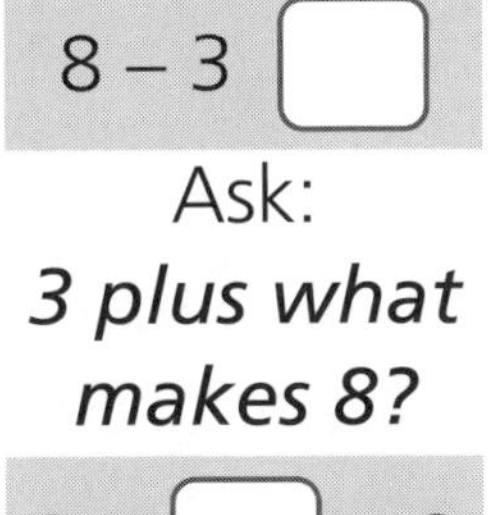

9 – 7 □

Ask:

7 plus what makes 9?

7 + □ = 9

You could also use the number line.

0 1 2 3 4 5 6 7 8 9 10 11 12 13 14 15 16 17 18 19 20

A	7 – 0 □	3 – 2 □	7 – 1 □	6 – 3 □	5 – 4 □
B	9 – 8 □	4 – 3 □	4 – 1 □	9 – 9 □	8 – 0 □
C	2 – 1 □	6 – 4 □	6 – 1 □	5 – 3 □	9 – 5 □
D	7 – 4 □	3 – 1 □	7 – 6 □	7 – 5 □	9 – 6 □
E	7 – 7 □	8 – 1 □	6 – 5 □	8 – 6 □	9 – 8 □
F	9 – 7 □	8 – 5 □	5 – 1 □	5 – 5 □	9 – 0 □
G	4 – 0 □	4 – 2 □	9 – 3 □	6 – 0 □	7 – 2 □
H	8 – 3 □	5 – 2 □	6 – 2 □	8 – 7 □	8 – 2 □
I	8 – 4 □	9 – 2 □	7 – 3 □	9 – 4 □	6 – 6 □
J	10 – 3 □	10 – 5 □	10 – 2 □	10 – 6 □	
K	10 – 8 □	10 – 7 □	10 – 4 □	10 – 9 □	

BLM 3 Subtraction facts to 20

14 – 6 ☐

Ask:
6 plus what makes 14?

6 + ☐ = 14

16 – 9 ☐

Ask:
9 plus what makes 16?

9 + ☐ = 16

You could also use the number line.

0 1 2 3 4 5 6 7 8 9 10 11 12 13 14 15 16 17 18 19 20

A	14 – 7 ☐	12 – 3 ☐	11 – 6 ☐	18 – 9 ☐
B	12 – 8 ☐	13 – 7 ☐	17 – 8 ☐	16 – 8 ☐
C	10 – 9 ☐	10 – 3 ☐	12 – 3 ☐	11 – 5 ☐
D	11 – 3 ☐	13 – 4 ☐	10 – 5 ☐	10 – 8 ☐
E	11 – 2 ☐	12 – 4 ☐	11 – 7 ☐	16 – 9 ☐
F	13 – 6 ☐	11 – 9 ☐	11 – 4 ☐	11 – 8 ☐
G	14 – 6 ☐	13 – 9 ☐	15 – 7 ☐	13 – 8 ☐
H	10 – 7 ☐	16 – 7 ☐	14 – 8 ☐	15 – 9 ☐
I	15 – 6 ☐	10 – 4 ☐	12 – 7 ☐	14 – 9 ☐
J	12 – 9 ☐	14 – 5 ☐	12 – 5 ☐	12 – 6 ☐
K	13 – 5 ☐	10 – 6 ☐	15 – 8 ☐	17 – 9 ☐

BLM 4 Skip counting / number chart

Skip count by 2. 2, 4, 6, 8, 10, 12, 14, 16, 18, 20 ...

Skip count by 10. 10, 20, 30, 40, 50, 60, 70, 80, 90, 100 ...

Skip count by 5. 5, 10, 15, 20, 25, 30, 35, 40, 45, 50 ...

Skip count by 4. 4, 8, 12, 16, 20, 24, 28, 32, 36, 40 ...

1	2	3	4	5	6	7	8	9	10
11	12	13	14	15	16	17	18	19	20
21	22	23	24	25	26	27	28	29	30
31	32	33	34	35	36	37	38	39	40
41	42	43	44	45	46	47	48	49	50
51	52	53	54	55	56	57	58	59	60
61	62	63	64	65	66	67	68	69	70
71	72	73	74	75	76	77	78	79	80
81	82	83	84	85	86	87	88	89	90
91	92	93	94	95	96	97	98	99	100
101	102	103	104	105	106	107	108	109	110
111	112	113	114	115	116	117	118	119	120
121	122	123	124	125	126	127	128	129	130